An Introduction to CATIA V5

Release 19

(A Hands-On Tutorial Approach)

Kirstie Plantenberg

University of Detroit Mercy

ISBN: 978-1-58503-534-2

SDC

PUBLICATIONS

Schroff Development Corporation

www.schroff.com

Schroff Development Corporation
P.O. Box 1334
Mission KS 66222
(913) 262-2664
www.schroff.com

Publisher: Stephen Schroff

PREFACE

An Introduction to CATIA (A Hands-On Tutorial Approach) is a collection of tutorials meant to familiarize the reader with CATIA's mechanical design workbenches. The reader is not required to have any previous CATIA knowledge. The workbenches covered in this book are; Sketcher, Part Design, Wireframe and Surface Design, Assembly Design, and Drafting. Preceding each tutorial is a brief description of the workbench, toolbars, and commands to be used and focused on within the tutorial. This book is not meant to be strictly a reference book. It is meant to enable the reader to get right into CATIA and start drawing. You will learn by doing. The author directs the reader to CATIA's Companion for in depth reference material and a more detailed description of the commands.

Changes from Release 17

For the most part, the commands and methods of constructing a part or an assembly within the *Mechanical Design* workbenches has remained the same. There are a few minor changes in some of the definition windows.

Occasionally an existing *Sketch* will disappear when a *Part Design* operation is applied to it such as *Pad*, *Pocket*, *Shaft*, etc.... This is an annoying problem, but easily overcome.

IMPORTANT! If you find that a sketch has disappeared, as described above, perform the following procedure.

1) Right click on the *Sketch* in the specification tree and select **Copy**.
2) Right click on the *Part Body* in the specification tree and select **Paste**.
3) The new *Sketch* should work.
4) It is also a good idea to delete the original *Sketch* and any erroneous *Sketches* that may have appeared.

Suggested time line for a 15 week course

CATIA Tutorials	CATIA Exercises	Week
CATIA Basics: Tutorials 1.1, 1.2 & 2.1	Exercise 2.1	Week 1
Sketcher: Tutorials 2.2, 2.3	Exercise 2.2, 2.3	Week 2
Sketcher: Tutorials 2.4, 2.5 & 2.6	Exercise 2.4, 2.5, 2.6	Week 3
Part Design: Tutorials 2.7, 3.1 & 3.2	Exercise 3.1, 3.2	Week 4
Part Design: Tutorials 3.3 & 3.4	Exercise 3.3	Week 5
Part Design: Tutorial 3.5	Exercise 3.4	Week 6
Part Design: Tutorials 3.6 & 3.7	Exercise 3.5	Week 7
Part Design: Tutorials 3.8 & 3.9		Week 8
Wireframe: Tutorial 4.1		Week 9
Wireframe: Tutorials 4.2 & 4.3	Exercise 4.1	Week 10
Assembly: Tutorial 5.1		Week 11
Assembly: Tutorial 5.2	Exercise 5.1	Week 12
Drafting: Tutorial 6.1	Exercise 6.1	Week 13

NOTE: One week is left available for spring break or computer mishaps. Another week is left available for the final.

ACKNOWLEDGMENTS

I would like to thank my mother, Phyllis Plantenberg, for painstakingly going over and performing each tutorial. This was an invaluable service enabling me to correct mistakes, misinterpretations, and vagueness. I would also like to thank the University of Detroit Mercy for providing the tools necessary to make this book possible.

TABLE OF CONTENTS

Chapter 1: CATIA BASICS

Introduction

Chapter 1 focuses on CATIA's interface and file management. The tutorials within this chapter are meant to help the reader navigate through CATIA's workbenches as well as to show how objects are selected.

!Note to UNIX users!

The tutorials within this book were designed for the Windows version of CATIA. There may be slight differences in how CATIA looks within a UNIX environment. One operational difference is this; in the UNIX version, to multiple select objects while in a command, the user must hold the CTRL key down. If the user does not do this, the current command window will disappear behind the document window.

Tutorials Contained in Chapter 1

- Tutorial 1.1: CATIA's interface and file management
- Tutorial 1.2: Selecting, Editing & Viewing Objects

NOTES:

Chapter 1: CATIA BASICS

Tutorial 1.1:
CATIA's Interface and file management

Featured Topics & Commands

Prerequisite Knowledge & Commands

- The windows operating system

CATIA's Interface

CATIA's main interface uses standard windows icons and commands such as New, Open, Save, Cut, Copy, and Paste. Therefore, the environment outside the drawing area should look very familiar. It is the environment inside the drawing area or document window that may take a while to get used to. Shown below are the main components of CATIA's user interface followed by a description of each area.

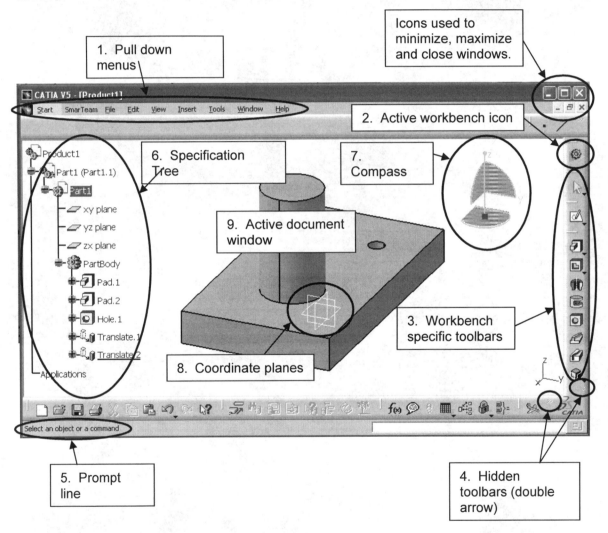

1. <u>Pull down menus:</u> All of the commands, options, and settings may be accessed in the pull down menus.
2. <u>Active workbench icon:</u> CATIA has several workbench environments. For example, the *Sketcher* workbench is used to sketch 2D profiles and the *Part Design* workbench is used to make solids from a sketch among other things. The active workbench icon shows the workbench you are currently working in.

3. <u>Workbench specific toolbars:</u> Each workbench has its own specific set of toolbars. However, some toolbars are common between workbenches.
4. <u>Hidden toolbars:</u> If there is not enough room to show all active toolbars, CATIA will hide some. You will know when some toolbars are hidden if a double arrow appears in one or more of the corners.
5. <u>Prompt line:</u> The prompt line tells the user what the program is expecting next. For example, if the *Circle* command is chosen, the prompt line will read *Select a point or click to define the circle center.* After the center point is selected it will read *Select a point or click to define the circle radius.*
6. <u>Specification tree:</u> The specification tree shows all the commands and sketches that were used to construct the part. The specification tree may be used to make changes to the part.
7. <u>Compass:</u> The compass may be used to rotate and translate the environment or part.
8. <u>Coordinate planes:</u> The default coordinate planes show the 3 mutually perpendicular Cartesian coordinate planes. 2D profiles may be sketched on these planes.
9. <u>Active document window:</u> CATIA allows multiple documents or parts to be open simultaneously. These documents can be arranged in the usual windows based placements such as <u>T</u>iled horizontally and vertically, <u>C</u>ascaded, etc...

Mechanical Design Workbenches

Listed below are the most commonly used Mechanical Design workbenches along with a brief summary of their uses. The figure shows all of CATIA's Product lines and standard Mechanical Design workbenches.

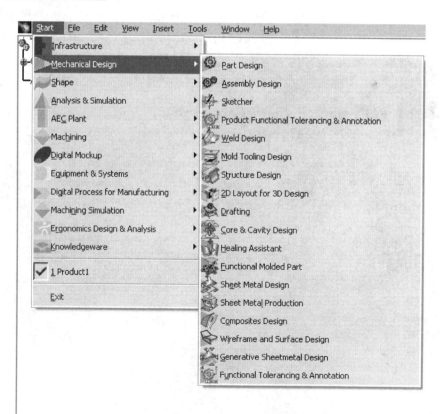

- <u>Part Design:</u> The Part Design workbench allows you to create a main solid from a sketch and then modify it by adding or removing material.

- <u>Sketcher:</u> The Sketcher workbench is used to create 2D profiles that may be used to create the initial solid, modify an existing solid, or used as a path or direction curve.

- <u>WireFrame & Surface Design:</u> The WireFrame & Surface workbench is used to create complex wire frame profiles and surfaces.

- <u>Assembly Design:</u> The Assembly Design workbench is used to assemble and constrain parts in an assembly.

- <u>Drafting:</u> The Drafting workbench is used to create 2D detailed and assembly drawings from 3D models.

Documents

When working on a part or assembly in CATIA you are working within a document. The type of document depends on the workbench you are using to create and modify the geometry and each has its own file extension.

- Assembly (.CATProduct)
- Part Design, Sketcher, WireFrame & Surface (.CATPart)
- Drafting (.CATDrawing)

Shown on the next page, are 3 document windows open simultaneously in CATIA. There is an assembly of the product called *Machine Vise*, a part (the *Base*) belonging to the product and a 2D drawing of the *Base*.

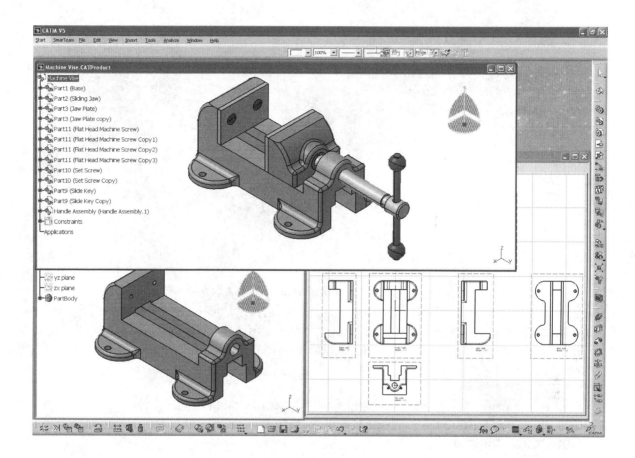

Tutorial 1.1 Start: Part Modeled

The part modeled in this tutorial is simple and is used to familiarize you with CATIA's interface.

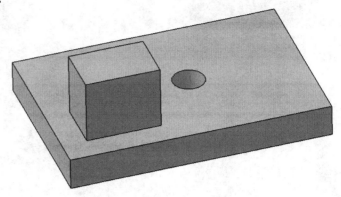

Section 1: CATIA's Workbenches

1) Launch CATIA by double clicking on the **CATIA V5** icon on your desktop. (UNIX users: Check with you system administrator for the CATIA startup procedure.) Two windows will appear the outer *CATIA V5* window and the inner *Product* window.

2) At the top pull down menu, select **Start** – **Mechanical Design** – **Part Design**. Workbenches may be accessed from the top pull down menu or from the workbench toolbar.

3) At the top pull down menu, select **File** – **New ….** In the *New* window, select **Part** and then **OK**. In the *New Part* window, name your part **Plate** and select **OK**. Maximize the *Part* window if it is not already maximized. Notice the specification tree. At the moment you have 3 coordinate planes and an empty PartBody.

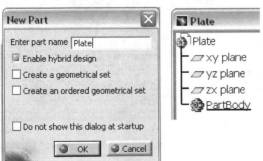

4) Create a list of favorite workbenches. At the top pull down menu, select **Tools** – **Customize.** This may take a while to load.

5) In the *Customize* window, select the **Start Menu** tab. Use the arrows to move the following workbenches from the *Available* area to the *Favorites* area: Assembly Design, Drafting, Part Design, Sketcher, and Wireframe and Surface Design. Then select the **Close** button.

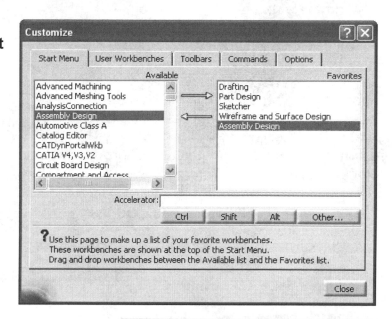

6) Select the **Part Design** workbench icon . A *Welcome to CATIA V5* window will appear that contains your favorite workbenches. This window will appear every time you start CATIA. **Close** the *Welcome to CATIA V5* window.

Section 2: The Specification Tree

2.1 – Specification Tree Basics

In the upper left corner of the *Part* window, there is a specification tree. This contains all the planes, sketches, operations, and constraints, used to create the part. The specification tree can be expanded and collapsed by clicking on the '+' and '-' symbols (not present at the moment). The most important feature of the specification tree is that it allows you to view and modify the individual components and operations used to create the part. The specification tree is arranged in a hierarchical manner. At the top of the tree is the Part, under the Part there are 3 coordinate planes and an empty PartBody. The tree's structure will change depending on the type of document you are editing.

1) In the specification tree, click on the **xy plane**. Notice that the xy coordinate plane in the middle of the screen highlights in orange. Repeat for each coordinate plane.

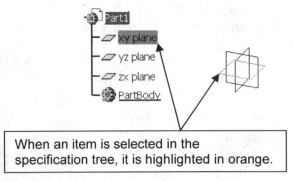

When an item is selected in the specification tree, it is highlighted in orange.

CATIA's interface allows you to reposition toolbars to suit your needs. Therefore, they may not always be located in their original position. By restoring the toolbars default position, locating specific icons throughout each tutorial is easier.

2) Restore the toolbars to their default positions. At the top pull down menu, select **Tools – Customize...**

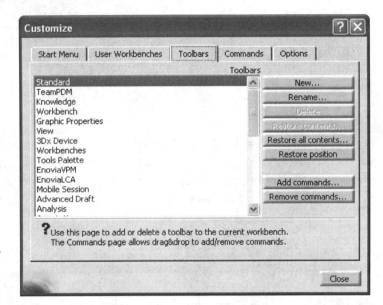

3) In the *Customize* window select the **Toolbars** tab and then select the **Restore all contents ....** In the *Restore all content* window, select **OK**. **Restore position** button. In the *Restore all toolbars* window select **OK** and then select **Close** in the *Customize* window. This procedure should be repeated any time you are unable to locate a specific toolbar.

4) Enter the *Sketcher* Workbench on the xy plane. Select the **xy plane** in the specification tree or the coordinate plane itself. Then select the **Sketcher** icon. This is usually located in the top right corner. In the *Sketcher* workbench a grid will appear and the environment will rotate such that you are viewing the xy plane.

5) Drag and drop the *User Selection Filter* and *Sketch Tools* toolbars to the top creating a top toolbar area. Place them in the approximate locations shown.

6) Locate the *Profile* toolbar in the right side toolbar area. Click on the bar at the top of this toolbar and drag it to the center of the screen. (If the *Profile* toolbar is not there, right click on the right side toolbar area and select Profile. A toolbar is active if there is a check mark next to it. UNIX users select from the top pull down menu, **View – Toolbars - Profile**)

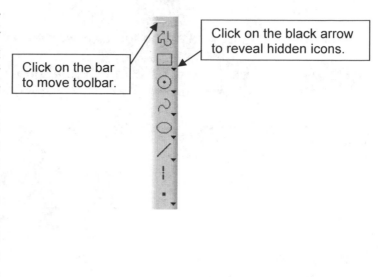

Click on the bar to move toolbar.

Click on the black arrow to reveal hidden icons.

7) Click on the black arrow under the Rectangle icon  to reveal the *Predefined Profile* toolbar.

8) Restore the toolbars default position as was done previously and drag the *User Selection Filter* and *Sketch Tools* toolbars to the top again.

9) Draw a Rectangle similar to the one shown. Use the grid as a guide. Select the **Rectangle** icon [ ]. It is usually located in the *Profile* toolbar. Notice that when you select the rectangle icon it turns orange. Down at the bottom of the screen the prompt line will read *Select or pick first point to create the rectangle*. Once you select the first point, the prompt line will read *Select or pick second point to create the rectangle*. The prompt line should be consulted whenever you are unsure what action to take next.

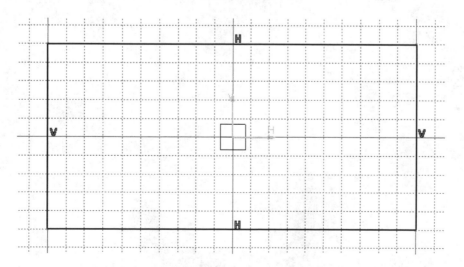

10) Exit the sketcher by selecting the **Exit workbench** icon .

11) Hit the **Esc** key twice or click on the graphics screen to deselect the rectangle. The rectangle should no longer be orange.

12) Click on the '+' symbol next to the *PartBody* to reveal *Sketch.1*.

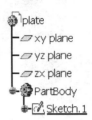

13) Pad the rectangle to a length of 20 mm. Select

the **Pad** icon . In the *Pad Definition* window, fill in the following fields;
- **Type**: **Dimension**
- **Length**: *20 mm*
- The Selection: field should read No selection. Activate the Selection: field by clicking on No selection and then click on the rectangle. The Selection: field should now read **Sketch.1**. Select **OK**. (UNIX users hold the CTRL key down when selecting the sketch.)

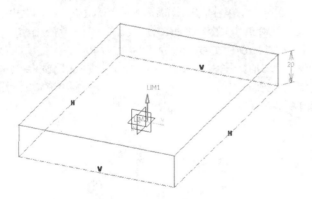

14) Create a 20 mm hole in the top of the block. Click on the **Hole** icon

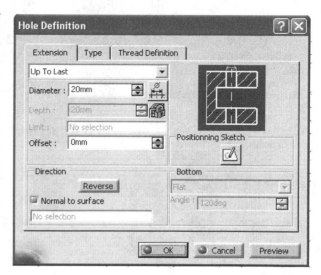

located in the right side toolbar area. Notice nothing happens. Read the prompt line. Select the top face of the block near the middle. In the *Hole Definition* window and **Extension** tab, fill in the following fields;

- Select **Up To Last** in the top pull down field,
- Diameter: *20 mm*.

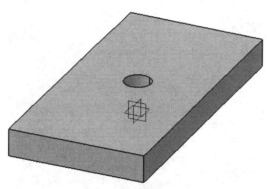

15) Expand your specification tree as shown. Notice that the PartBody in the specification tree is no longer empty. It contains a main solid (Pad.1) and a hole (Hole.1). Notice that each operation is based on a separate sketch.

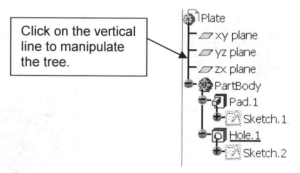

Click on the vertical line to manipulate the tree.

2.2 – Manipulating the Specification Tree

The specification tree may be moved, zoomed or hidden.

1) Click on one of the vertical connecting lines in the tree. Your part should turn a dark grey this indicates that the tree is active. (The specification tree may also be activated and deactivated by using **Shift + F3**.)

2) Move the tree to the center of the screen by clicking on the vertical line again and dragging it.

3) Zoom in on your tree by clicking and holding your middle mouse button, clicking your right mouse button once, and moving your mouse up and down.

4) Move the tree back to its original location in the upper left corner and zoom it to its normal size.

5) Click on the vertical connecting line again to deactivate the tree.

6) Hit the **F3** key on your keyboard to hide the tree and then hit it again to show the tree.

Section 3: File Management

1) At the top pull down menu select **File – Save**. A Save As window will appear because this is the first save. Select the file location and name your file *T1-1* then select the **Save** button. The extension for a Part Design document is *CATPart*.

2) Enter the Sketcher on the top face of the block. Select the

   Sketcher icon and then select the top face of the block.

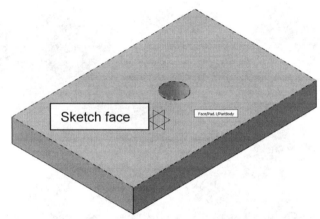

Sketch face

3) Draw a **Rectangle**

   in the approximate location and of approximate size as shown.

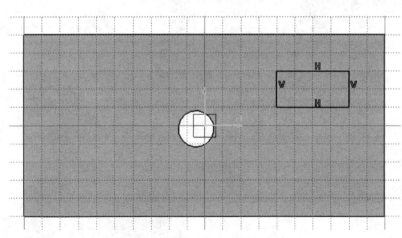

4) Exit the sketcher by selecting the **Exit workbench** icon .

5) **Pad** 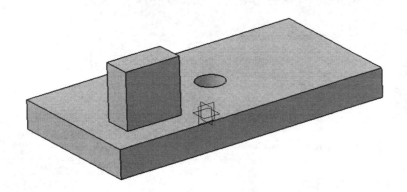 the rectangle to a length of **40 mm**.

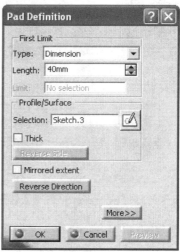

6) At the top pull down menu select **F**ile – **S**ave.

7) At the top pull down menu select **F**ile – **C**lose. If a *Close* window appears, Select **N**o.

8) At the top pull down menu select **F**ile – **N**ew From... Check the **Show Preview** box and click on **T1-1.CATPart**. Select the **Open** button.

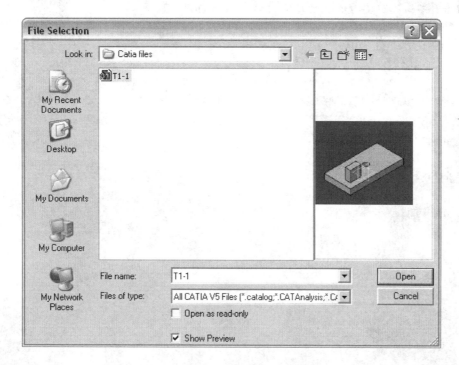

9) Notice that CATIA automatically renames your file **T1-1_1**. Save the new file.

Section 4: Using the Compass

The compass is used to translate and rotate the 3D environment and all of its components or an individual part.

1) Click and hold one of the straight lines of the compass and move your mouse. Notice that both the part and the coordinate planes translate.

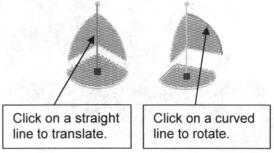

Click on a straight line to translate.	Click on a curved line to rotate.

2) Click and hold one of the curved lines of the compass and move your mouse. Notice that the part and the coordinate planes both rotate.

3) Click and hold on the top point of the compass and move your mouse. This allows you to rotate freely with no coordinate constraints.

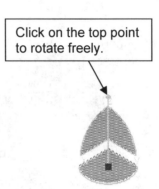

Click on the top point to rotate freely.

4) Attach the compass to your part. Click and drag the red square at the base of the compass and move it to the top of your part and release. The compass is attached to the part if it turns green. If it is not green or it snaps back to it original position, move it again until it attaches to the part.

5) Repeat steps 1) and 2) and notice that the coordinate planes do not move. Only the part translates or rotates. If any warning windows popup informing you of constraint violations, click **OK** or **Yes**.

6) From the top pull down menu select **V**iew – **R**eset Compass.

7) Close without saving.

Chapter 1: CATIA BASICS

Tutorial 1.2:
Selecting, Editing & Viewing Objects

Featured Topics & Commands

Prerequisite Knowledge & Commands

- Entering workbenches

Simple Selection

The left mouse button is used to select an object. You can select an object or feature on the physical part or you can click on the corresponding feature in the specification tree. To select multiple items hold the control key down while selecting. The selected item(s) will be highlighted in orange in both the tree and the part.

The Selection Toolbar

There are several ways to select a feature. You can simply select the feature using your left mouse button or you can use one of the available traps. The *Selection* toolbar has 7 different trap options and is located near the top of the right side toolbar area. Reading the toolbar from left to right the commands are;

- <u>Select:</u> This allows you to select features by using your left mouse button. To multiple select you need to hold the control key down. This option is always active.
- <u>Selection trap above Geometry:</u> This option allows you to start the trap on a specific element and not on an empty space as with the other trap options.
- <u>Rectangle Selection Trap:</u> The selection trap allows you to create a window and select all features that are completely enclosed in that window.
- <u>Intersecting Rectangle Selection Trap:</u> The intersecting trap uses a window to select all items that are enclosed in or touching the window boundaries.
- <u>Polygon Selection Trap:</u> The polygon trap is similar to the selection trap but it enables you to create a window that is irregular.
- <u>Free Hand Selection Trap:</u> This selection trap allows you to create a free hand line. Everything that crosses this line is selected.
- <u>Outside Trap Rectangle Selection:</u> The outside trap selection selects everything that is outside the window.
- <u>Intersecting Outside Rectangle Trap Selection:</u> The intersecting outside trap selection selects everything that is outside and intersecting the window.

The View Toolbar

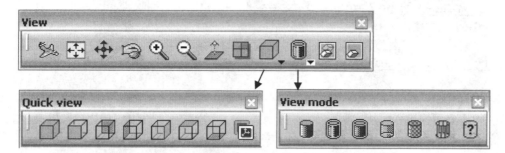

The *View* toolbar contains commands that allow you to manipulate (pan, rotate, zoom) your part and contains different viewing options. Reading the toolbar from left to right the options are;

- Fly Mode: Allows you to dynamically fly around the part. To use the fly mode you must be in a perspective view point (<u>V</u>iew – Rendering St<u>y</u>le – <u>P</u>erspective)
- Fit All In: This zooms your part out or in to fit the available window space.
- Pan: Allows you to move your part around the screen without zooming.
- Rotate: Allows you to rotate your part to view different sides.
- Zoom In: Makes your part look bigger.
- Zoom Out: Makes your part look smaller.
- Normal View: It rotates the part to show you a 2D view of the object relative to a selected face.
- Create Multi-view: This command allows you to create four different independently controlled views of the part and/or sketch.
- Quick view toolbar: Allows you to view the part from the following different perspectives.
  - Isometric view
  - Front view
  - Back view
  - Left view
  - Right view
  - Top view
  - Bottom view
  - Named views

- View mode toolbar: Enables you to view the part in the following different modes.
  - Shading (SHD)
  - Shading with Edges
  - Shading with Edges without Smooth Edges
  - Shading with Edges and Hidden Edges
  - Shading with Material
  - Wireframe (NHR)
  - Customize View Parameters

- <u>Hide/Show:</u> Allows you to hide and show different features of a part.
- <u>Swap visible space:</u> This command hides everything that is showing and shows everything that is hidden.

The Graphics Properties Toolbar

The *Graphic Properties* toolbar allows you to change the fill color, line thickness and type, and transparency of an object among other things. The options in the *Graphic Properties* toolbar from left the right are;

- <u>Fill Color:</u> Allows you to change the color of a part.
- <u>Transparency:</u> You can make a part transparent so that you can see parts or features behind it.
- <u>Line Weight:</u> You can change the line thickness of a sketch.
- <u>Line Type:</u> You can change the line type of a sketch.
- <u>Point Symbol:</u> There are several different choices for point style.
- <u>Painter:</u> This allows you to match the color of one feature to the color of another.
- <u>Wizard:</u> The Wizard command allows you to view the graphic properties of different features of your part.

Tutorial 1.2 Start: Part Modeled

The part modeled in this tutorial is shown on the right. It will be used to illustrate how to edit, manipulate, view and change the graphic properties of a part.

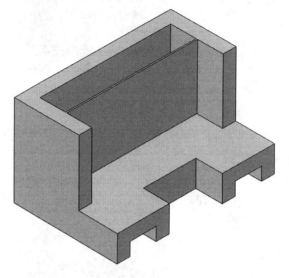

Section 1: Selecting profiles and features

1) Launch **CATIA V5**, enter the **Part Design** workbench, enter a **New Part** and name your part **Seat**. Save your part as **T1-2.CATPart**.

2) Enter the Sketcher workbench on the xy plane. Select the **xy plane** in the specification tree or the coordinate plane it self then select the **Sketcher** icon .

3) Turn the snap to grid option on. Select the **Snap to Point** icon. If the icon is orange, the snap is on, if it is blue, the snap is off. The *Snap to Point* icon is located in the *Sketch Tools* toolbar. This toolbar should be located at the top toolbar area where it was moved. If this toolbar is not active or you can't locate it, go to **Tools – Customize...** – **Toolbar** tab – **Restore all Content** and **Restore Position**.

4) Select the **Profile** icon and draw the sketch shown. The profile command allows you to draw several continuous lines connected together. Count the grid spacing to determine the sketch size.

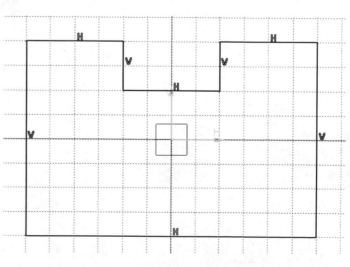

5) Click on the screen somewhere away from the sketch to deselect the profile. Nothing should be highlighted in orange.

6) Click on the left vertical line of the profile. Notice it turns orange. Click on the bottom line in the profile. Notice that the bottom line is orange, but the vertical line is no longer orange. Hold the control key and click on the right side vertical line of the profile. Notice that the bottom line remains selected.

7) Deselect all. Click off of the profile or hit the **Esc** key twice.

8) Bring the **Select toolbar** out from underneath the **Select** icon and move it so that it stands alone.

9) In the *Select* toolbar, select the **Rectangle Selection Trap** icon. Draw the selection window shown by clicking and holding the mouse button and moving to the opposite corner. Notice that the only the lines that where entirely enclosed in the window are selected.

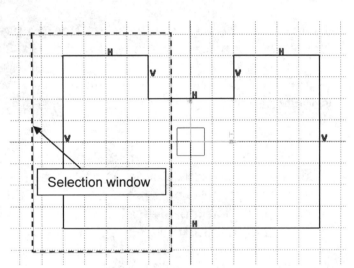

Selection window

10) Deselect all.

11) In the *Select* toolbar, select the **Intersecting Rectangle Selection Trap** icon. Draw a selection window like the previous window. Notice that this time all items that where enclosed in and touching the window are selected.

12) Deselect all.

13) In the *Select* toolbar, select the **Polygon Selection Trap** icon. Draw the selection window shown clicking the mouse at positions 1 to 4 in order. Double click at point 4 to end the window selection.

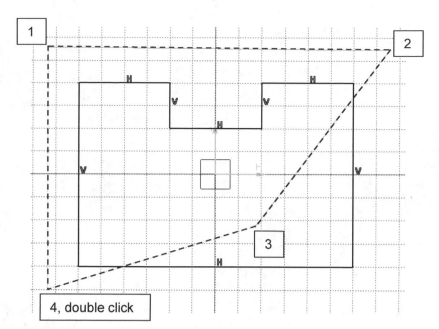

14) Deselect all.

15) Try the other trap commands and determine what they do.

16) Use the **Rectangle Selection Trap** to select the entire sketch.

17) Hold the **control** key and deselect the bottom line by clicking on it.

18) Close the **Select toolbar**.

19) Exit the sketcher by selecting the **Exit workbench** icon.

20) Pad the profile to a length of 20 mm. Select the **Pad** icon. In the *Pad Definition* window fill in the following fields;
- Type: **Dimension**
- Length: **20 mm**
- Selection: **Sketch.1**

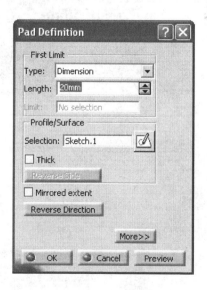

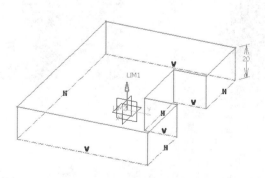

Section 2: Undo/Redo

1) Enter the Sketcher workbench on the top face of the part. Select the

   Sketcher icon 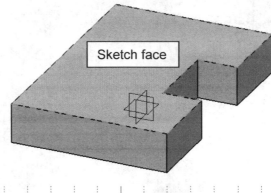 and then select the top face of the part.

   Sketch face

2) Draw the following **Profile**
   .

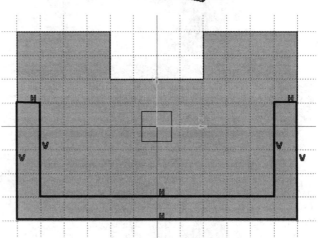

3) **Exit** the Sketcher and **Pad** the sketch to a length of *50 mm*.

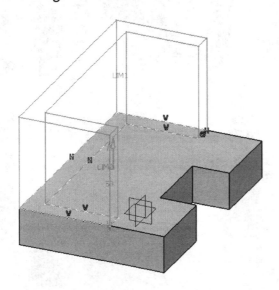

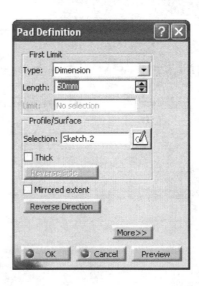

4) Save your part.

5) Enter the **Sketcher** on the **zx plane**.

6) Draw the **Line** shown in white.

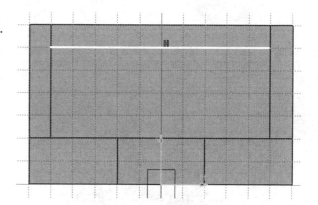

7) **Exit** the Sketcher .

8) Select the **Stiffener** icon. This is located in the *Sketch Based Features* toolbar. It may be stacked under the **Solid Combine** icon. Create a stiffener using **Sketch.3** that is *1 mm* thick and is **From the Side**.

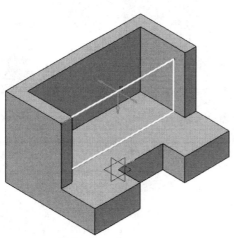

9) Enter the **Sketcher** on the front face of the part.

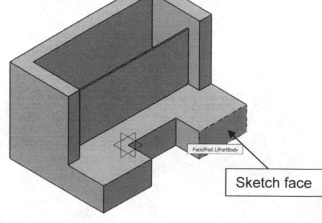

Sketch face

10) Draw 2 **Rectangles** as shown.

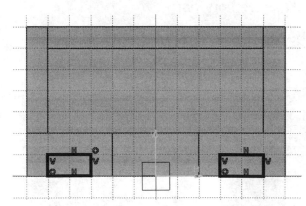

11) **Exit** the Sketcher .

12) Create pockets using the two rectangles. The Pocket command removes material. Select the **Pocket** icon and use the **Up to Last** option (this means through all).

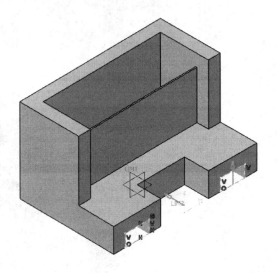

13) Expand your specification tree to the sketch level as shown.

14) Select **Stiffener.1** in the specification tree. Notice that this feature is highlighted in the specification tree as well as in the part.

15) Right click on **Stiffener.1** and select **Delete**. Select **OK** in the *Delete* window.

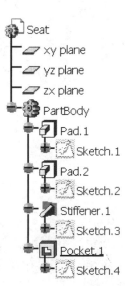

16) At the top pull down menu, select **Edit – Undo.** Alternatively, you can select the **Undo** 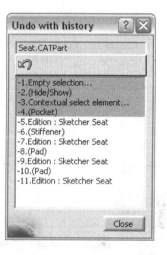 icon, or use the keyboard shortcut **CTRL + Z**.

17) Right click on **Sketch.3** (the sketch associated with the stiffener) and select **Hide/Show**. Now we can see the sketch used to create the stiffener.

18) Select the **Undo with History** icon. This is stacked under the *Undo* icon. Your undo history should look similar to the one shown. There may be some extra steps depending on what you have done. Select **(Pocket).** This will select all commands performed before the Pocket as well as the Pocket command. Select the **Undo** icon and then **Close**.

Undo with history ? X

Seat.CATPart

-1.Empty selection...
-2.(Hide/Show)
-3.Contextual select element...
-4.(Pocket)
-5.Edition : Sketcher Seat
-6.(Stiffener)
-7.Edition : Sketcher Seat
-8.(Pad)
-9.Edition : Sketcher Seat
-10.(Pad)
-11.Edition : Sketcher Seat

Close

19) At the top pull down menu, select **Edit – Repeat**. Alternatively you can select the **Redo** icon or use the keyboard shortcut **CTRL + Y** to get the Pocket back.

Section 3: Editing features using the compass

1) Deselect all.

2) Click and drag the red square of the compass to the top surface of your original pad as shown. Notice that *Pad.1* highlights in the specification tree.

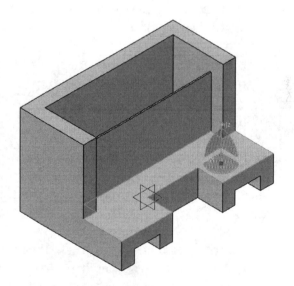

3) In the specification tree, click on **Stiffener.1**. Now the compass is attached to the stiffener instead of the pad.

4) Right click on the compass and select **Edit...**

5) In the *Parameters for Compass Manipulation* window, enter **10 mm** in each of the Translation increment fields (U, V & W) and select **Close**. By doing this you are setting the minimum increment that the compass is allowed to move.

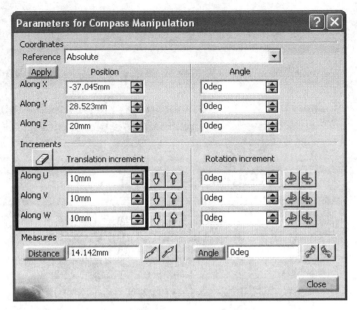

6) Move the compass one increment in the Y or V direction and one increment in the Z or W direction. When the *Move Warning* window appears select **Yes** or **OK**. (Hint: To translate the compass, click and drag on one of the straight axis lines.) When complete your part should look like the figure shown.

7) At the top pull down menu, select **View** – **Reset Compass**.

8) Save your part.

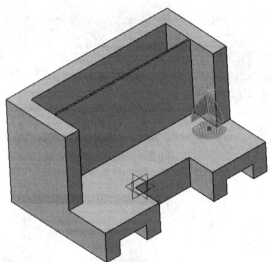

Section 4: Scanning the design process

This allows you to view the part's history step by step in order to see the design process.

1) At the top pull down menu, select **Edit** – **Scan or Define In Work Object...**

2) In the *Scan* window, select the **First** button, then scroll through the design process using the **Next** button. When complete, select the **Exit** button.

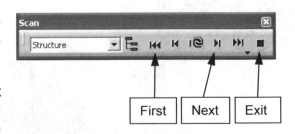

Section 5: Viewing your part

The part can be manipulated (Pan, Rotate, Zoom) using the icons in the *View* toolbar or by using the mouse buttons. Given below is a description of how to use a three button mouse to manipulate your part.

- <u>Pan</u> = Hold the middle mouse button and move the mouse.
- <u>Zoom</u> = Hold the middle mouse button, click and release the right mouse button and move the mouse.
- <u>Rotate</u> = Hold the middle mouse button, click and hold the right or left mouse button and move the mouse.

1) Use the icons in the *View* toolbar to **Pan** , **Rotate** , and **Zoom** your part. This toolbar is located in the bottom toolbar area. It may be hidden in the bottom right corner.

2) Use your mouse buttons to Pan, Rotate and Zoom your part.

3) Select the **Fit All In** icon.

4) View your object from the **Top view** , then the **Right view** , and then the **Isometric view** .

5) Select different view modes and notice the differences . Go back to **Shaded with Edges** when finished.

Section 6: Graphic Properties

Parts in CATIA can be assigned different colors. This is very helpful especially when viewing an assembly.

1) In your specification tree, right click on **PartBody** and select **Properties**.

2) In the *Properties* window, select the *Graphic* tab. Under the Fill Color field select the color **red** and select **Apply**. Notice that the entire part changes to red. Then, under the Edges Color field select the color **green** and select **OK**.

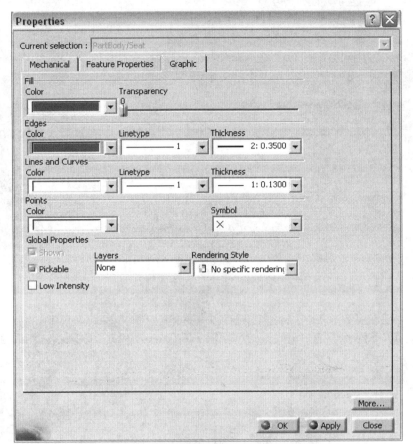

3) Deselect all.

4) In your specification tree, right click on **Stiffener.1** and select **Properties**. Set the fill color to **tan**.

5) At the top pull down menu, select **View** – **Toolbars** – **Graphic Properties**. The *Graphics Properties* toolbar should appear.

6) In the specification tree, select **Pad.2** and change its fill color to **cyan** (light blue) using the *Graphic Properties* toolbar.

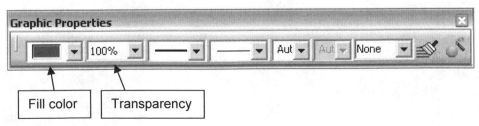

7) In the specification tree, select **Stiffener.1** and change its Transparency to **50%** using the *Graphic Properties* toolbar.

8) Using the **control** key, pick both surfaces that were used to sketch the pocket shapes as shown in the figure and change there color to **cyan** (light blue).

(Problem? Did your pockets disappear. If so, you did not complete the *Scan or Define in Work Object* process.)

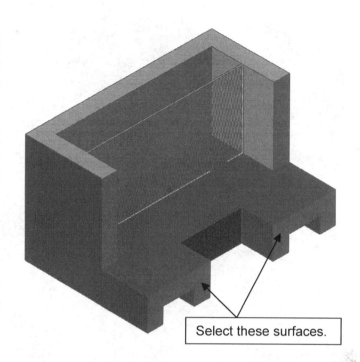

Select these surfaces.

9) Deselect all.

10) Hold **control key** down and select **Pad.1** in the specification tree. Still holding the control key, select the **Painter** 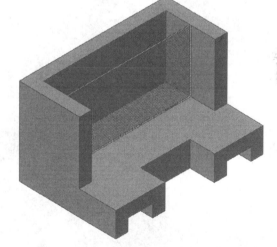 icon and select **Pad.2**. This procedure matches the color of Pad.1 to Pad.2.

11) Using the **Painter** icon, change the color of the **Pocket.1** to make the color of **Pad.1**.

12) Save your part.

NOTES:

Chapter 2: SKETCHER

Introduction

Chapter 2 focuses on CATIA's *Sketcher* workbench. The reader will learn how to sketch and constrain very simple to very complex 2D profiles.

Tutorials Contained in Chapter 2

- Tutorial 2.1: Sketch Work Modes
- Tutorial 2.2: Simple Profiles & Constraints
- Tutorial 2.3: Advanced Profiles & Sketch Analysis
- Tutorial 2.4: Modifying Geometries & Relimitations
- Tutorial 2.5: Axes & Transformations
- Tutorial 2.6: Operations on 3D Geometries & Sketch planes
- Tutorial 2.7: Points & Splines

NOTES:

Chapter 2: SKETCHER

Tutorial 2.1: Sketch Work Modes

Featured Topics & Commands

Prerequisite Knowledge & Commands

- Entering workbenches
- Entering and exiting the *Sketcher* workbench
- Drawing simple profiles
- Simple Pads and Pockets

The Sketcher Workbench

The *Sketcher* workbench contains a set of tools that help you create and constrain 2D geometries. Solid features such as pads, pockets and shafts are created or modified using these 2D profiles. You can access the *Sketcher* workbench in various ways. Two simple ways are by using the top pull down menu (<u>S</u>tart – <u>M</u>echanical Design – <u>S</u>ketcher), or by selecting the *Sketcher* icon. When you enter the *Sketcher*, CATIA requires that you choose a plane to sketch on. You can choose this plane either before or after you select the *Sketcher* icon. To exit the sketcher, select the *Exit Workbench* icon.

The *Sketcher* workbench contains the following standard workbench specific toolbars.

- <u>Profile toolbar:</u> The commands located in this toolbar allow you to create simple geometries (rectangle, circle, line, etc...) and more complex geometries (profile, spline, etc...).

- <u>Operation toolbar:</u> Once a profile has been created, it can be modified using commands such as trim, mirror, chamfer, and other commands located in the *Operation* toolbar.

- <u>Constraint toolbar:</u> Profiles may be constrained with dimensional (distances, angles, etc...) or geometrical (tangent, parallel, etc...) constraints using the commands located in the *Constraint* toolbar.

- <u>Sketch tools toolbar:</u> The commands in this toolbar allow you to work in different modes which make sketching easier.

- <u>User Selection Filter toolbar:</u> Allows you to activate different selection filters.

- Visualization toolbar: Allows you to, among other things to cut the part by the sketch plane and choose lighting effects and other factors that influence how the part is visualized.

- Tools toolbar: Allows you to, among other things, to analyze a sketch for problems, and create a datum.

The Sketch tools Toolbar

The *Sketch tools* toolbar contains icons that activate and deactivate different work modes. These work modes assist you in drawing 2D profiles. Reading from left to right, the toolbar contains the following work modes; (Each work mode is active if the icon is orange and inactive if it is blue.)

- Grid: This command turns the sketcher grid on and off.
- Snap to Point: If active, your cursor will snap to the intersections of the grid lines.

- Construction / Standard Elements: You can draw two different types of elements in CATIA a *standard* element and a *construction* element. A standard element (solid line type) will be created when the icon is inactive (blue). Standard elements are used to create a feature in the *Part Design* workbench. A construction element (dashed line type) will be created when the icon is active (orange). Construction elements are used to help construct your sketch, but will not be used to create features.
- Geometric Constraints: When active, geometric constraints will automatically be applied such as tangencies, coincidences, parallelisms, etc...
- Dimensional Constraints: When active, dimensional constraints will automatically be applied when corners (fillets) or chamfers are created, or when quantities are entered in the value field. The value field is a place where dimensions such as line length and angle are manually entered.

Tutorial 2.1 Start: Part Modeled

The part modeled in this tutorial is shown below. The part is constructed with the assistance of different work modes.

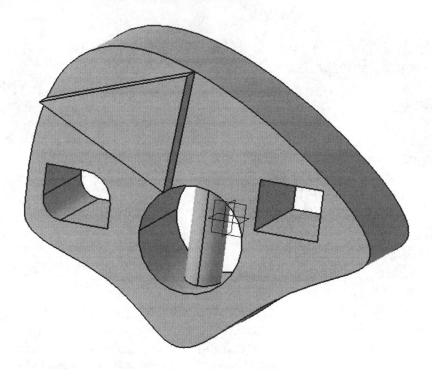

Section 1: Using Snap to Point

1) Open a **New Part** drawing and name the part *Spline Shape*.

2) Save your drawing as *T2-1.CATPart*.

3) Enter the **Sketcher** on the **yz plane**.

4) Restore the default positions of the toolbars (**Tools** – **Customize...** – **Toolbars** tab – **Restore all contents...** & **Restore position**.) Move the *Sketch Tools* toolbar and the *User Selection Filter* toolbar to the top toolbar area.

5) Set your grid spacing to 100 mm. At the top pull down menu, select **Tools –**
Options... In the *Options* window, expand the **Mechanical Design** portions
of the left side navigation tree and select **Sketcher**. In the Grid section,
activate the following checkboxes and fill in the following fields:
- Activate Display, Snap to point, and Allow Distortions.
- Set your Primary spacing and Graduations to H: ***100 mm*** and ***20***,
  and V: ***100 mm*** and ***10***.

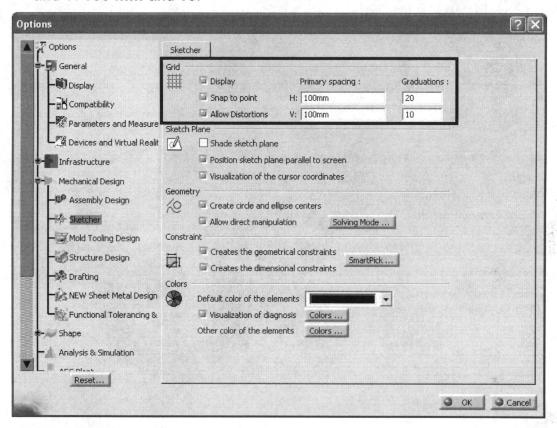

6) Select the **Spline** icon located in the *Profile* toolbar in the right side
toolbar area. This is <u>not</u> the *Curve Filter* icon located in the *User
Selection Filter* toolbar

7) In your *Sketch Tools* toolbar, activate your **Grid** icon and your **Snap to
Point** icon. It should be orange (active). Move your cursor around the
screen. Note that it snaps to the intersections of the grid. Deactivate the
Snap to Point icon by clicking on it and turning it back to blue. Move
your cursor around the screen and notice the difference.

8) Reactivate the **Snap to Point** icon and draw the spline shown. Select each point (indicated by a number in a square) in order from 1 to 7, double clicking at the last point to end the spline command.

9) Edit the spline by double clicking on any portion of it.

10) In the *Spline Definition* window, select **CtrlPoint.7**, then activate the **Tangency** option, and select **OK**. Notice that the last point is now tangent to the first point. (Problem? If the tangency is not working, go back and make sure that your points are located in the correct locations.)

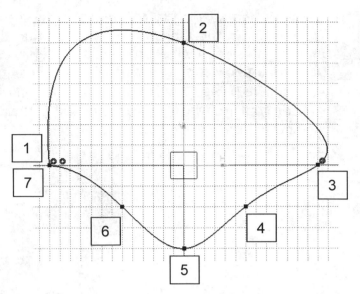

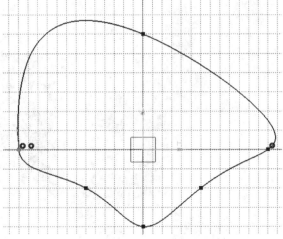

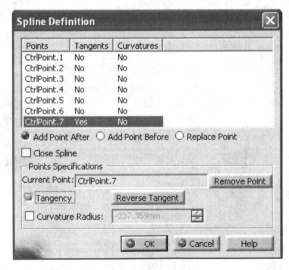

11) Draw a **Circle** inside the spline as shown.

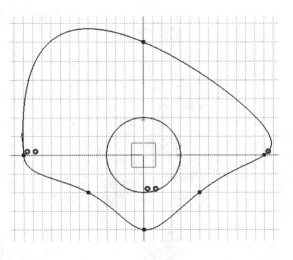

12) **Exit** the Sketcher 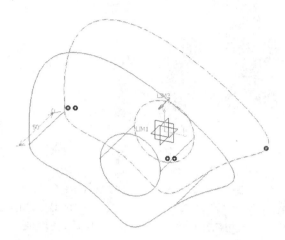 and **Pad** the sketch to a length of *50 mm*.

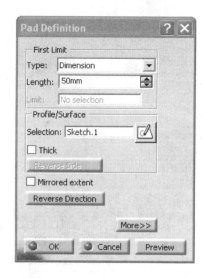

13) **Save** your drawing.

Section 2: Using construction elements.

1) Deselect all.

2) Enter the **Sketcher** on the front face of the part.

3) Activate the **Construction / Standard Elements** icon. It should be orange.

4) Deselect all. Hit the **Esc** key twice.

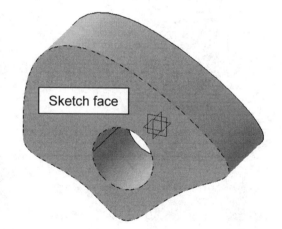

Sketch face

5) Project an outline of the part onto the sketch plane. Select the **Project 3D Elements** icon then select the face of the part. This icon is located in the *Operations* toolbar near the bottom of the right side toolbar area. It may be hidden in the bottom right corner.

6) Deselect all. The projection should now be yellow (this means it is associated with the part and will change with the part) and dashed (this means it is a construction element).

7) Deactivate your **Grid** ▦, **Snap to Point** ▦, and **Construction / Standard Elements** ◌ icons.

8) Activate your **Geometrical constraints** ⧉ and **Dimensional constraints** ⌐ icons. They should be orange.

9) Using the **Profile** ⌂ command to draw a triangle that looks like the one shown. The points of the triangle should lie on the projected construction element. You will know when you are on the projection when a symbol of two concentric circles appears, and you will know when you are snapped to the endpoint of the start point when a symbol of two concentric circles appears and the inner one is filled.

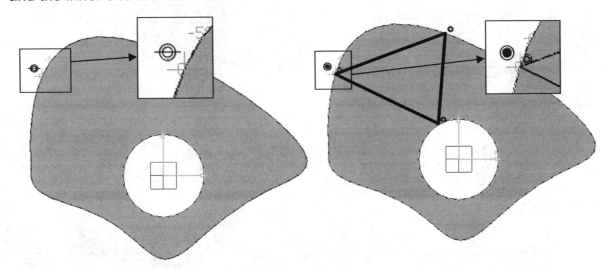

10) **Exit** the Sketcher ⬆ and **Pad** ⬈ the sketch to a length of **10 mm**.

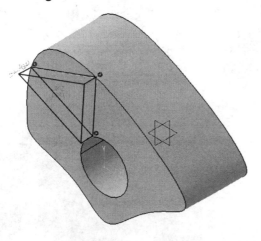

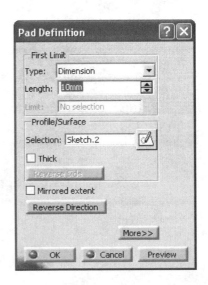

Section 3: Geometrical and Dimensional Constraints

1) Deselect all.

2) Enter the **Sketcher** <img> on the front large face of the part.

3) Your **Geometrical Constraints** <img> icon should be active. It should be orange.

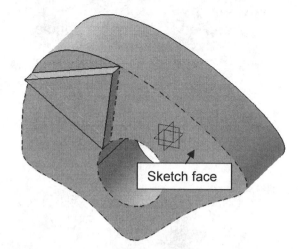

Sketch face

4) At the top pull down window, select **Tools – Options – Sketcher**. Under the Constraint portions of the window, select **SmartPick...** The *SmartPick* window shows all the geometrical constraints that will be created automatically. These constraints may be turn on and off depending on your design/sketch needs. **Close** both the *Smart Pick* and *Options* windows.

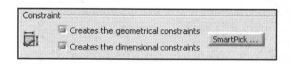

5) Draw a **Rectangle** <img> to the right of the hole as shown. Notice that geometric constraints (H = horizontal, V = Vertical) are automatically applied.

6) Deactivate the **Geometrical Constraints** <img> icon and draw a **Rectangle** <img> to the left of the hole as shown. Notice that no geometric constraints are made.

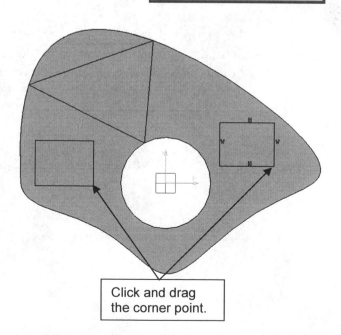

Click and drag the corner point.

7) For each rectangle, click on one of the points defining a corner and move it using the mouse (see figure on the previous page). Notice the difference between the two. This is due to the horizontal and vertical constraints that were applied to the one rectangle.

8) Undo (**CTRL + Z**) the moves until the original rectangles are back.

9) **Exit** the Sketcher and **Pocket** the sketch using the **Up to last** option.

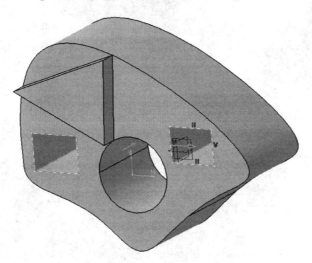

10) Expand the specification tree to the sketch level.

11) **Save** your drawing.

12) Edit Sketch.3 (the sketch associated with the pocket). In the specification tree, double click on **Sketch.3**, or right click on it and select **S̲ketch.3 object - E̲dit**. You will automatically enter the sketcher on the sketch plane used to create this sketch.

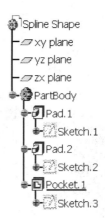

13) Your **Dimensional Constraint** icon should be active. It should be orange.

14) Select the **Corner** ⌒ icon, select the bottom left corner point of the left rectangle, move your mouse up and to the right, and click. A corner or fillet will be created. The corner icon is located in the *Operations* toolbar near the bottom of the right side toolbar area. The corner/fillet may also be created by selecting the two lines that create the corner. Notice that a dimension is automatically created.

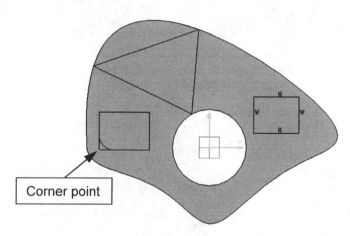

Corner point

15) Deactivate the **Dimensional Constraint** ⌶ icon. It should be blue. Create a **Corner** ⌒ in the upper right corner of the same rectangle. Notice that this time no dimensional constraint was created.

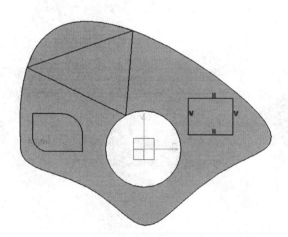

16) **Exit** the Sketcher ⬆. We have changed the sketch used to create the pocket. Notice that the pocket is automatically updated to reflect these changes.

17) **Save** your drawing.

Section 4: Cutting the part by the sketch plane.

Sometimes it is necessary to sketch inside the part. The *Cut Part by Sketch Plane* command allows you to see inside the part and makes it easier to draw and constrain your sketch.

1) Deselect all.

2) Enter the **Sketcher** on the **xy plane**.

3) Select the **Isometric View** icon. This icon is located in the bottom toolbar area.

4) Select the **Cut Part by Sketch**

   Plane icon located in the bottom toolbar area. The part in now cut by the xy plane (the sketch plane).

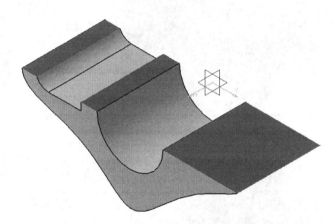

5) Select the **Top view** icon

   and draw a **Circle** in the middle of the hole as shown in the figure.

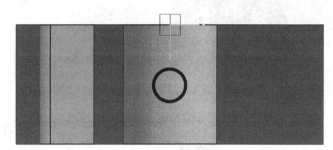

6) **Exit** the Sketcher.

7) Select the **Pad** 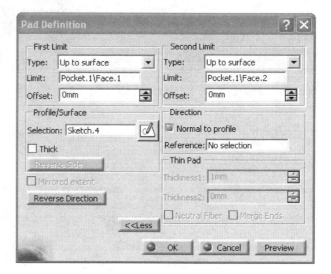 icon and then select the **More>>** button. Fill in the following fields for both the First and Second Limits;

- Type: **Up to surface**
- Limit: Select the inner circumference of the hole
- Selection: **Sketch.4** (the circle).

Select **Preview** to see if the Pad will be applied correctly, and then **OK**.

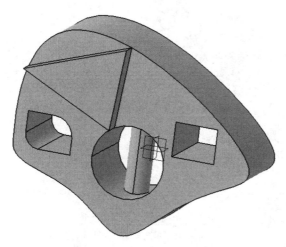

8) **Save** your drawing.

NOTES:

Chapter 2: SKETCHER

Tutorial 2.2: Simple Profiles & Constraints

Featured Topics & Commands

Prerequisite Knowledge & Commands

- Entering workbenches
- Entering and exiting the *Sketcher* workbench
- Simple Pads
- Work modes (Sketch tools toolbar)

Profile toolbar

The *Profile* toolbar contains 2D geometry commands. These geometries range from the very simple (point, rectangle, etc...) to the very complex (splines, conics, etc...). The *Profile* toolbar contains many sub-toolbars. Most of these sub-toolbars contain different options for creating the same geometry. For example, you can create a simple line, a line defined by two tangent points, or a line that is perpendicular to a surface.

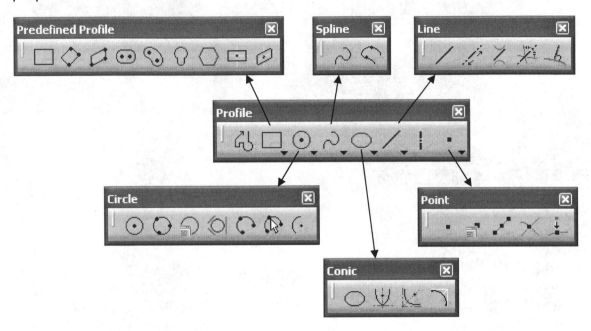

Profile toolbar

Reading from left to right, the *Profile* toolbar contain the following commands.

- Profile: This command allows you to create a continuous set of lines and arcs connected together.
- Rectangle / *Predefined Profile* toolbar: The default top command is *rectangle*. Stacked underneath are several different commands used to create predefined geometries.
- Circle / *Circle* toolbar: The default top command is *circle*. Stacked underneath are several different options for creating circles and arcs.
- Spline / *Spline* toolbar: The default top command is *spline* which is a curved line created by connecting a series of points.
- Ellipse / *Conic* toolbar: The default top command is *ellipse*. Stacked underneath are commands to create different conic shapes such as a hyperbola.
- Line / *Line* toolbar: The default top command is *line*. Stacked underneath are several different options for creating lines.

- Axis: An axis is used in conjunction with commands like mirror and shaft (revolve). It defines symmetry. It is a construction element so it does not become a physical part of your feature.
- Point / *Point* toolbar: The default top command is *point*. Stacked underneath are several different options for creating points.

Predefined Profile toolbar

Predefined profiles are frequently used geometries. CATIA makes these profiles available for easy creation which speeds up drawing time. Reading from left to right, the *Predefined Profile* toolbar contains the following commands.

- Rectangle: The *rectangle* is defined by two corner points. The sides of the rectangle are always horizontal and vertical.
- Oriented Rectangle: The *oriented rectangle* is defined by three corner points. This allows you to create a rectangle whose sides are at an angle to the horizontal.
- Parallelogram: The *parallelogram* is defined by three corner points.
- Elongated Hole: The elongated hole or slot is defined by two points and a radius.
- Cylindrical Elongated Hole: The *cylindrical elongated hole* is defined by a cylindrical radius, two points and a radius.
- Keyhole Profile: The *keyhole profile* is defined by two center points and two radii.
- Hexagon: The *hexagon* is defined by a center point and the radius of an inscribed circle.
- Centered Rectangle: The *centered rectangle* is defined by a center point and a corner point.
- Centered Parallelogram: The *centered parallelogram* is defined by a center point (defined by two intersecting lines) and a corner point.

Circle toolbar

The *Circle* toolbar contains several different ways of creating circles and arcs. Reading from left to right, the *Circle* toolbar contains the following commands.

- Circle: A *circle* is defined by a center point and a radius.
- Three Point Circle: The *three point circle* command allows you to create a circle using three circumferential points.
- Circle Using Coordinates: The *circle using coordinates* command allows you to create a circle by entering the coordinates for the center point and radius in a *Circle Definition* window.

- Tri-Tangent Circle: The *tri-tangent circle* command allows you to create a circle whose circumference is tangent to three chosen lines.
- Three Point Arc: The *three point arc* command allows you to create an arc defined by three circumferential points.
- Three Point Arc Starting With Limits: The *three point arc starting with limits* allows you to create an arc using a start, end, and midpoint.
- Arc: The *arc* command allows you to create an arc defined by a center point, and a circumferential start and end point.

Spline toolbar

Reading from left to right, the *Spline* toolbar contains the following commands.

- Spline: A spline is a curved profile defined by three or more points. The tangency and curvature radius at each point may be specified.

- Connect: The connect command connects two points or profiles with a spline.

Conic toolbar

Reading from left to right, the *Conic* toolbar contains the following commands.

- Ellipse: The ellipse is defined by a center point and major and minor axis points.
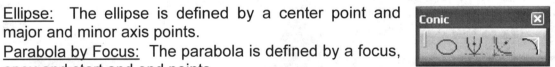
- Parabola by Focus: The parabola is defined by a focus, apex and start and end points.
- Hyperbola by Focus: The hyperbola is defined by a focus, center point, apex and start and end points.
- Conic: There are several different methods that can be used to create conic curves. These methods give you a lot of flexibility when creating the above three types of curves.

Line toolbar

The *Line* toolbar contains several different ways of creating lines. Reading from left to right, the *Line* toolbar contains the following commands.

- Line: A line is defined by two points.

- Infinite Line: Creates infinite lines that are horizontal, vertical or defined by two points.
- Bi-Tangent Line: Creates a line whose endpoints are tangent to two other elements.
- Bisecting Line: Creates an infinite line that bisects the angle created by two other lines.

- <u>Line Normal to Curve:</u> This command allows you to create a line that starts anywhere and ends normal or perpendicular to another element.

<u>Point toolbar</u>

The *Point* toolbar contains several different ways of creating points. Reading from left to right, the *Point* toolbar contains the following commands.

- <u>Point by Clicking:</u> Creates a point by clicking the left mouse button.
- <u>Point by using Coordinates:</u> Creates a point at a specified coordinate point.
- <u>Equidistant Points:</u> Creates equidistant points along a predefined path curve.
- <u>Intersection Point:</u> Creates a point at the intersection of two different elements.
- <u>Projection Point:</u> Projects a point of one element onto another.

Constraint toolbar

Constraints can either be dimensional or geometrical. Dimensional constraints are used to constrain the length of an element, the radius or diameter of an arc or circle, and the distance or angle between elements. Geometrical constraints are used to constrain the orientation of one element relative to another. For example, two elements may be constrained to be perpendicular to each other. Other common geometrical constraints include parallel, tangent, coincident, concentric, etc... Reading from left to right:

- <u>Constraints Defined in Dialoged Box:</u> Creates geometrical and dimensional constraints between two elements.
- <u>Constraint:</u> Creates dimensional constraints.
  - o <u>Contact Constraint:</u> Creates a contact constraint between two elements.
- <u>Fix Together:</u> The fix together command groups individual entities together.
  - o <u>Auto Constraint:</u> Automatically creates dimensional constraints.
- <u>Animate Constraint:</u> Animates a dimensional constraint between to limits.
- <u>Edit Multi-Constraint:</u> This command allows you to edit all your sketch constraints in a single window.

Selecting icons

When an icon is selected, it turns orange indicating that it is active. If the icon is activated with a single mouse click, the icon will turn back to blue (deactivated) when the operation is complete. If the icon is activated with a double mouse click, it will remain active until another command is chosen or if the Esc key is hit twice.

Tutorial 2.2 Start: Part Modeled

The part modeled in this tutorial is shown on the right. This part will be created using simple profiles, circles, arcs, lines, and hexagons. The geometries are constrained to conform to certain dimensional (lengths) and geometrical constraints (tangent, perpendicular, etc...).

Section 1: Creating circles.

(Hint: If you get confused about how to apply the different commands that are used in this tutorial, read the prompt line for additional help.)

1) Open a **New...** part and name your part *Post*.

2) **Save** your drawing as *T2-2.CATPart*.

3) Enter the **Sketcher** 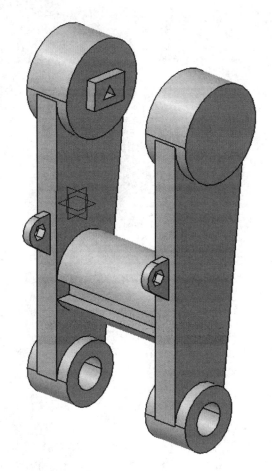 on the **zx plane**.

4) Set your grid spacing to be **100 mm** with **10** graduations, activate the *Snap to point*, and activate the *geometrical* and *dimensional* constraints. (**Tools** – **Options...**)

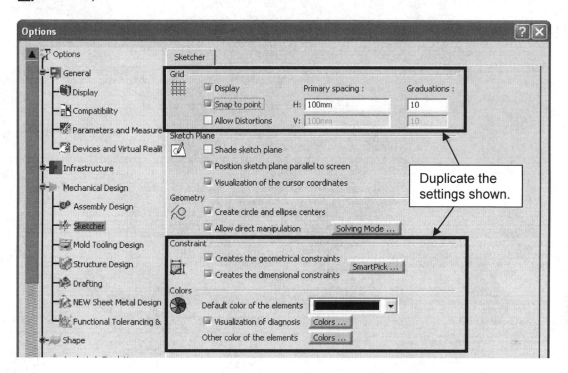

5) Pull out the **Circle** subtoolbar .

6) Double click on the **Circle**  icon and draw the circles shown.

7) **Exit** the Sketcher 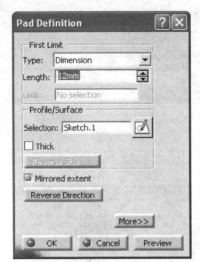 and **Pad** the sketch to *12 mm* on each side (**Mirrored extent**). Notice that the inner circle at the bottom becomes a hole.

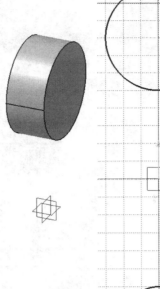

Section 2: Creating dimensional constraints.

1) Expand your specification tree to the sketch level.

2) Edit Sketch.1. To edit a sketch you can double click on the sketch name in the specification tree, or you can *right click* on the name select **Sketch.1 - Edit**. CATIA automatically takes you into the sketcher on the plane used to create Sketch.1.

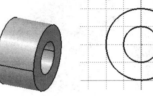

3) Double click on the **Constraints** icon.

4) Select the border of the upper circle, pull the dimension out and click your left mouse button to place the dimension. Repeat for the two bottom circles.

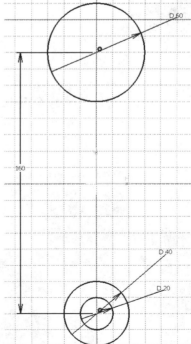

5) Select the center point of the upper circle, then the center point of the lower circles, pull the dimension out and click.

6) Double click on the **D20** dimension. In the *Constraint Definition* window, change the diameter from 20 to **16 mm**.

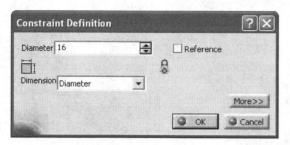

7) In a similar fashion, change the other dimensions to the values shown in the figure.

8) **Exit** the Sketcher 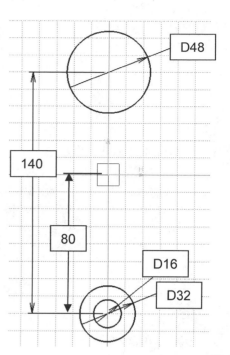 and deselect all. Notice that the part automatically updates to the new sketch dimensions.

Section 3: Creating lines.

1) Deselect all.

2) Enter the **Sketcher** on the **zx plane**.

3) Deactivate the **Snap to Point** icon.

4) Project the two outer circles of the part onto the sketch plane as Standard elements. Double click on the **Project 3D Elements** icon. This icon is located in the lower half of the right side toolbar area. Select the outer edges of the two cylinders.

5) Pull out the *line* toolbar .

6) Pull out the **Relimitations** toolbar located in the *Operation* toolbar.

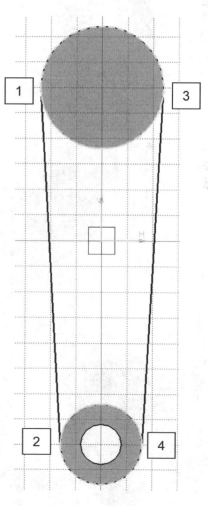

7) Double click on the **Bi-Tangent Line** icon. Draw two tangent lines by selecting the points, in order, as indicated on the figure.

8) Double click on the **Quick trim** icon. Select the outer portion of the projected circles. Notice that the trimmed projection turns into a construction element (dashed).

9) **Exit** the Sketcher and **Pad** the sketch to **6 mm** on each side (**Mirrored extent**).

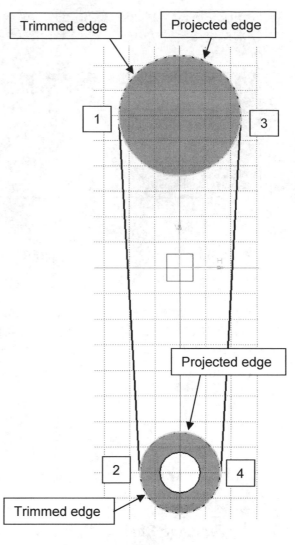

Trimmed edge

Projected edge

1

3

Projected edge

2

4

Trimmed edge

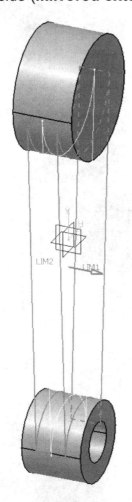

10) **Save** your drawing.

11) Enter the **Sketcher** on the **zx plane**.

12) Activate the **Construction/Standard Element** icon (it should be orange).

13) Select the **Project 3D Elements** icon and then project the left line of the part as shown in the figure. The projected line should be dashed.

14) Activate your **Snap to Point** icon.

15) Draw a line that starts at point 1 (see fig.) and ends normal/perpendicular to projected line using the **Line Normal to Curve** icon.

16) Deactivate your **Snap to Point** icon.

17) Draw a **Line** from point 1 to point 2.

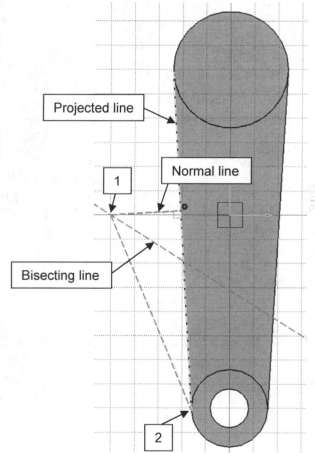

18) Draw a line that bisects the previous 2 lines using the **Bisecting Line** icon. Read the prompt line for directions.

19) Deselect all.

20) Deactivate the **Construction/Standard Element** icon (it should be blue now).

21) Draw a circle that is tangent to the projected line, normal line and bisecting line using the **Tri-Tangent Circle** 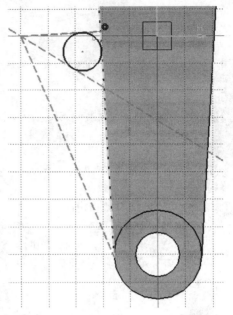 icon. Read the prompt line for directions.

22) Zoom in on the circle.

23) Using **Profile** , draw the three additional lines shown in the figure. When creating the line that touches the circle, both the construction line and the circle should turn orange before the point is selected.

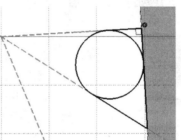

24) Use the **Quick Trim** command to trim off the inside portion of the circle as shown. You will have to apply the quick trim operation twice.

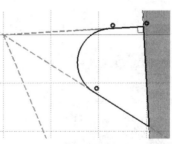

25) Draw a **Hexagon** that has the same center as the circle/arc and is the approximate size shown in the figure. The *Hexagon* icon is usually stacked under the *Rectangle* icon. (Your hexagon will contain many constraints that are not shown in the figure.)

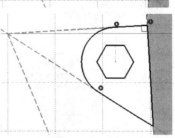

26) Deselect all.

27) Apply a dimensional **Constraint** 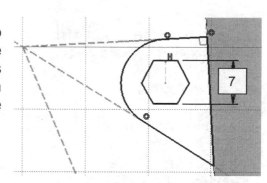 to the distance between the flats of the hexagon as shown. To create this constraint, select the top line and then the bottom line. Double click on the dimension and change its value to **7 mm**.

28) **Exit** the Sketcher and **Pad** the sketch to a length of **2 mm** on each side (**Mirrored extent**).

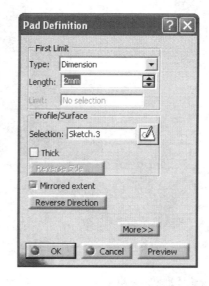

Section 4: Creating geometrical constraints.

1) Enter the **Sketcher** on the flat face of the large cylinder.

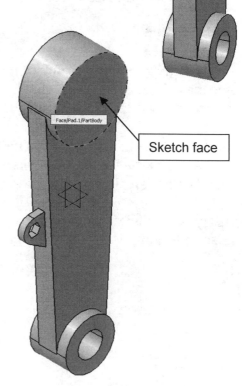

Sketch face

2) Deactivate the **Geometrical Constraint** icon (it should be blue). This will allow you to create profiles with no automatically applied constraints.

3) On the face of the large cylinder, draw the **Profile** shown. No geometrical constraints should be indicated.

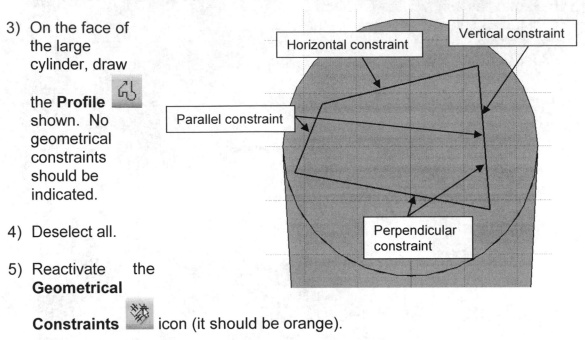

Horizontal constraint

Vertical constraint

Parallel constraint

Perpendicular constraint

4) Deselect all.

5) Reactivate the **Geometrical Constraints** icon (it should be orange).

6) Apply a vertical constraint to the right line of the profile by right clicking on it and selecting **L̲ine.? object – V̲ertical**.

7) Apply a horizontal constraint to the top line using a similar procedure.

8) Deselect all.

9) Apply a perpendicular constraint between the right and bottom line of the profile. Hold the CTRL key down and select the left and bottom lines. Select the **Constraints Defined in Dialog Box** icon. In the *Constraint Definition* window, check the box next to **Perpendicular** and then select **OK.**

10) Apply a parallel constraint between the left and right lines of the profile in a similar way.

11) Apply **Constraints** to the rectangle and change their values to the values shown in the figure.

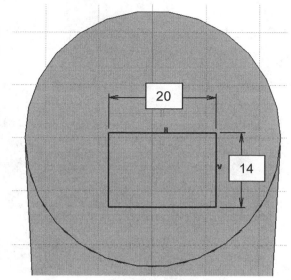

12) Apply the additional dimensional constraints shown in order to position the rectangle. Select the

Constraints icon, then the circumference of the circle and then the appropriate side of the rectangle. Notice that once all the constraints are applied, the rectangle turns green indicating that it is fully constrained. If it did not turn green make sure the *Visualization of diagnosis* is activated in the *Options* window. (Tools – Options...)

13) Draw the triangle shown using the

Profile command. When drawing the triangle make sure that the top point is aligned with the origin ( ) and the bottom line is horizontal (H).

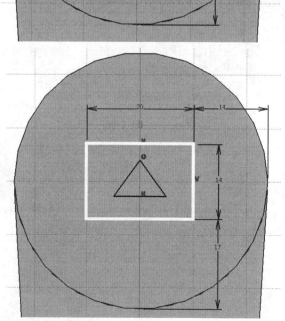

14) Constrain the vertical height of the triangle to be **6 mm**. Select the **Constraints** 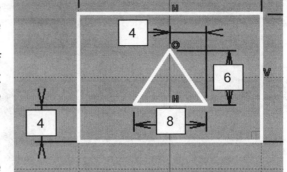 icon, select one of the angled lines of the triangle, right click and select **Vertical Measure Direction** and place the dimension.

15) **Constrain** the rest of the triangle as shown.

16) **Exit** the Sketcher and **Pad** the sketch to a length of **5 mm**.

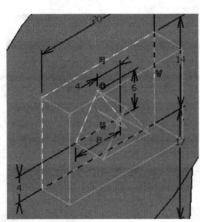

(Problem? If your sketch disappeared, *Copy* and *Paste* the sketch as described in the preface.)

17) **Save** your drawing.

Section 5: Creating arcs.

1) Enter the **Sketcher** on the front face of the middle section.

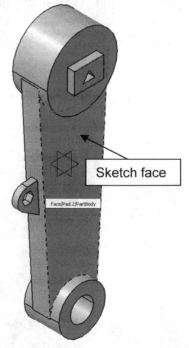

Sketch face

Face/Pad.2/PartBody

2) Activate the **Construction/Standard Element** 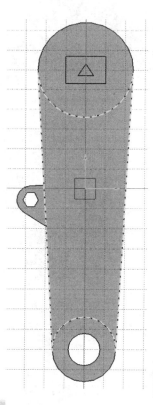 icon.

3) Select the **Project 3D Elements** icon and then project the front face of the middle section.

4) Deselect all.

5) Deactivate the **Construction/Standard Element** icon.

6) Activate your **Snap to Point** icon.

7) Draw the profile shown. Use the **Three Point Arc** command to create the bottom arc, the **Arc** command to create the top arc. The *Arc* icons are stacked under the *Circle* icon. For assistance in creating the arcs, read the prompt line at the bottom of the graphics screen.

Use the **Profile** command to create the connecting lines.

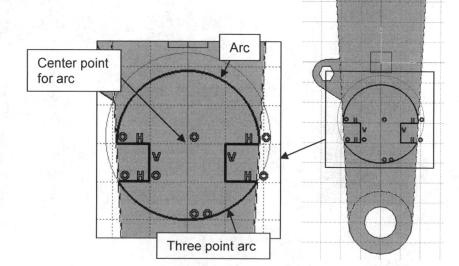

Center point for arc

Arc

Three point arc

8) **Exit** the Sketcher 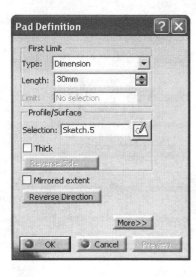 and **Pad** the sketch to a length of *30 mm.*

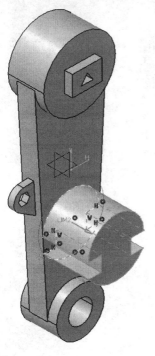

9) Deselect all.

10) Mirror the entire solid. Select the **Mirror** icon in the *Transformation Features* toolbar. Select the mirror element/face. In the *Mirror Definition* window select **OK**.

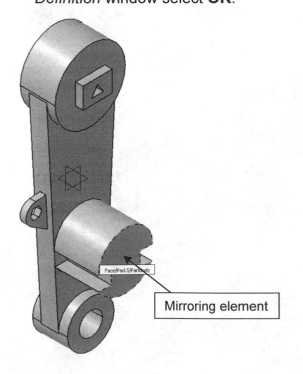

Mirroring element

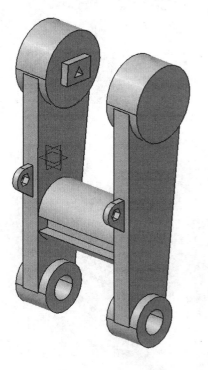

Chapter 2: SKETCHER

Tutorial 2.3: Advanced Profiles & Sketch Analysis

Featured Topics & Commands

Prerequisite Knowledge & Commands

- Entering workbenches
- Entering and exiting the *Sketcher* workbench
- Simple profiles
- Simple pads and pockets
- Applying dimensional and geometric constraints
- Work modes (*Sketch tools* toolbar)

Open and Closed Profiles

A profile is a series of geometric elements (points, lines, arcs, etc...) connected together. Profiles are used to create or modify solids. If they are not constructed properly, CATIA will not allow you to perform certain operations in the *Part Design* workbench. Most operations such as *Pad* and *Pocket*, with some exceptions, require that a profile be closed and not self intersecting.

- Open profile: A profile is open when the first and last elements of the series are not connected.

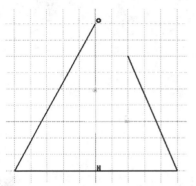

- Closed profile: A profile is closed when the last element connects up with the first element in the series.

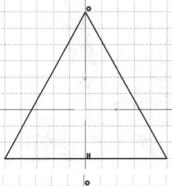

- Self intersecting profile: A self intersecting profile has at least two elements crossing each other.

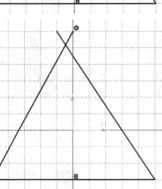

- Inner profiles: Inner profiles are closed profiles inside other closed profiles. The operation performed on an inner profile is opposite to the operation performed on the outer profile.

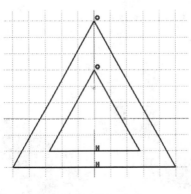

For example, if the outer profile is padded then the inner profile is pocketed.

The Value Field

When sketching in the *Sketcher* workbench, a value field appears that allows you to enter exact coordinates, lengths, angles and radii. The value field attaches itself to the *Sketch tools* toolbar. For example, consider the value field when the *Profile* command is active. For the first point of the profile, you can enter a horizontal and vertical coordinate in the value field. The 'Tab' key is used to access the H: and V: value fields. The 'Enter' key should not be used when entering values in the value field.

If you are drawing a line, the second point of the profile can be defined by a coordinate point (H,V) or a length and an angle (L,A).

For an arc, the second point is defined by a coordinate point (H,V) and the radius (R) of the arc may also be entered.

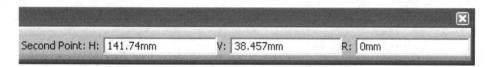

Each 2D geometry (Line, Rectangle, Elongated Hole, Arc, etc...) has its own specific value field that allows you to control the geometric characteristics of that geometry.

Sketch Status

The sketch status tells you if your sketch is under-constrained, over-constrained, or iso-constrained. Without constraints (geometrical and dimensional) a sketch can be moved freely just by clicking and dragging on an element. This could cause problems. If an element within the sketch is moved, it affects the solid that the sketch supports and any part that is related to the solid through assembly constraints. Therefore, it is important to completely constrain each sketch. If a sketch is completely constrained, it turns green.

- Under-constrained: A sketch is under-constrained when it is not completely constrained. In simple terms, there is not enough given information to duplicate the sketch exactly.

- <u>Over-constrained:</u> A sketch is over-constrained when there are duplicate dimensions given. The dimensions will turn purple when it is over-constrained. The duplicate dimension may be deleted or converted into a reference dimension.

- <u>Iso-constrained:</u> A sketch is iso-constrained when the sketch is completely constrained and there are enough dimensions and constraints given to duplicate the sketch exactly.

- <u>Inconsistent constraints:</u> Inconsistent constraints are constraints that conflict. For example, a parallel and perpendicular constraint applied to the same two lines is inconsistent or impossible. If there are inconsistent constraints present, the part will turn red.

Sketch analysis and sketch solving status

Within the *Tools* toolbar there is a *2D Analysis* subtoolbar containing commands that will help determine the sketch status and analyze sketch problems.

- <u>Sketch Solving Status icon:</u> This command will tell you if your sketch is under, over, or iso-constrained.

- <u>Sketch Analysis icon:</u> This command analyzes your sketch to determine problem areas. For example, it will identify any place where your sketch is open or intersecting, and it will identify duplicate and inconsistent constraints.

Tutorial 2.3 Start: Part Modeled

The part modeled in this tutorial is shown on the right. The main or outer profile is created using the help of the value field. The profiles used to modify the main pad are constructed in such a way as to illustrate the usefulness of solving for sketch status and analyzing sketches.

Section 1: Using the value field.

1) Enter a **New Part** drawing and name your part **Wedge**.

2) **Save** your drawing as **T2-3.CATPart**.

3) Enter the **Sketcher** on the **yz plane**.

4) Enter your *Options* window (**Tools – Options...**). In the *Sketcher* section, activate the grid's *Display* and *Snap to Point* options, and set the snap spacing for both directions to be: Primary spacing = **100 mm,** Graduations = **10 mm**. In the *Constraint* section, activate both the *Dimensional* and *Geometrical constraints*.

5) Select the **Profile** icon.

6) Position your cursor at **-70,70** and click your left mouse button. Notice that a value field appears attached to your *Sketch tools* toolbar. Some of the value field may not be visible. If this is the case, move the toolbar so that all parts are visible.

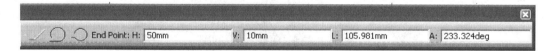

End Point: H: 50mm   V: 10mm   L: 105.981mm   A: 233.324deg

7) Hit the **TAB** key until the **L:** field is activated.   Type **140** in the **L:** field and hit the **TAB** key (do not hit the enter key).

8) Move your cursor around the screen. Notice that the angle of the line changes, but the length is constrained to be 140 mm and does not change.

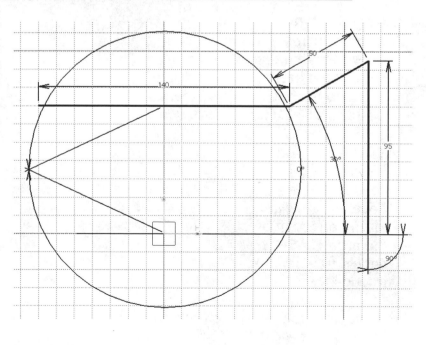

9) Hit the **TAB** key until the A: field is activated, type **0**, and hit the **TAB** key. Notice that dimensions are automatically applied for every value that was typed in the value field.

(Having trouble? If your value field is not retaining the values that you type, move the cursor off the graphics screen and try again. If the cursor moves on the graphics screen, the values in the value field change to match the cursor position.)

10) Use the same procedure to make the next two lines. First line: L = **50** and A = **30**, Second line: L = **95** and A = **270**.

11) In the value field, click on the **Tangent arc** icon. Notice that the field options change.

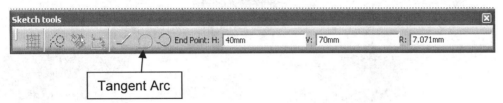

Tangent Arc

12) Hit the **TAB** key until the R: field is activated. Type **60**, and hit the **TAB** key.

13) Move your cursor around the graphics screen. Notice that the end point of the arc changes but not the radius.

14) Move your mouse until a vertical constraint line appears and click for the endpoint of the arc. The arc should travel through 90°.

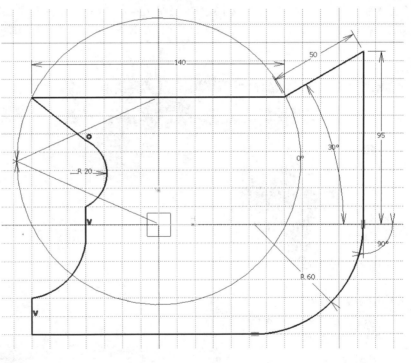

15) Using the value field, create two more lines: H = **-70**, V = **-60**, and then H = **-70**, V = **-40**.

16) In the value field, click on the **Three point arc** icon. Type the following values for the arcs second point: H = *-50*, V = *-30*.

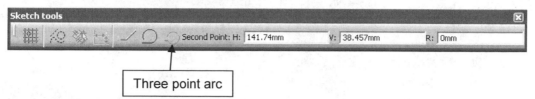

Three point arc

17) Move your mouse around the graphics screen and notice that the radius of the arc changes.

18) Type the following values for the arcs third point: H = *-40*, V = *-10*.

19) Create a **Line** (select this in the value field) whose endpoint is H = *-40*, V = *10*.

20) Create a **Three point arc** whose second point is H = *-30*, V = *20* and whose radius is R = *20*.

21) Move your mouse around the graphics screen. Notice that radius does not change. Move the mouse until a vertical constraint with the last line drawn is created and click.

22) Snap to profile's start point.

(Notice that every value that was entered into the value field has a dimension associated with it.)

23) **Exit** the *Sketcher* and **Pad** the sketch to a length of *50 mm*.

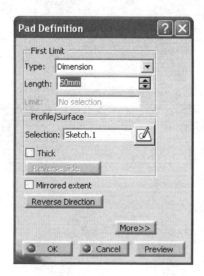

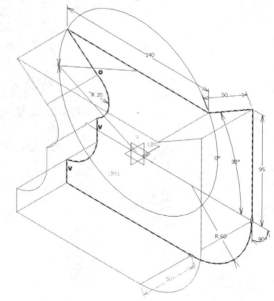

Section 2: Sketch Analysis

1) **Save** your drawing.

2) Enter the **Sketcher** 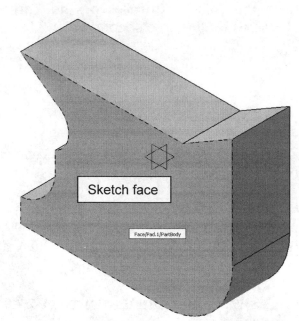 on the left face of the part.

3) Draw the **Profile** shown in the figure. Double click at the last point to end the *Profile* command.

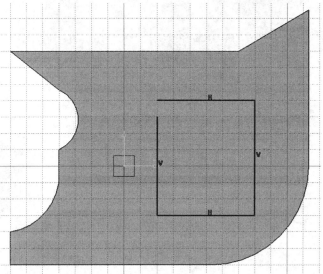

4) Select the **Sketch Analysis** icon. This is located in the *Tools* toolbar and usually found in the bottom toolbar area. The *Sketch Analysis* icon is usually stacked under the *Solve Sketch Status* icon.

5) A *Sketch Analysis* window will appear. Select the **Geometry** tab. In the Detailed information area, it should say that your profile is open. Notice that on your sketch two blue circles appear that identify the open section.

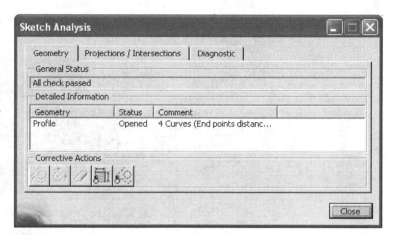

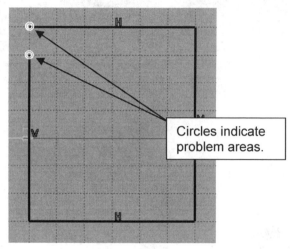

Circles indicate problem areas.

6) **Close** the *Sketch Analysis* window.

7) Draw a **Line** that starts at the end of the open vertical line and crosses the top horizontal line of the sketch.

8) **Exit** the *Sketcher* and try to apply a **Pocket** to the sketch. An error window will appear. This is because the profile is self intersecting. Select **OK** in the error window and **Cancel** the *Pocket Definition* window.

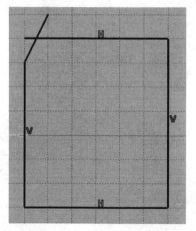

9) Edit Sketch.2 (the most recent sketch). This may be done by double clicking on the sketch in the tree or on the part, or right clicking on Sketch.2 in the specification tree and selecting Sketch.2 object – Edit.

10) Select the **Sketch Analysis** icon.

11) A *Sketch Analysis* window will appear. Select the **Geometry** tab. In the Detailed information area, it should say that your profile is autocrossing or self intersecting. Notice that on your sketch two blue circles appear that identify the problem area.

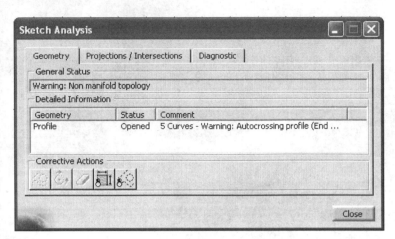

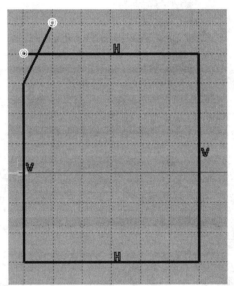

12) **Close** the *Sketch Analysis* window.

13) Select the **H** constraint of the top horizontal line and hit the **Delete** key. Then drag the lines left endpoint up to coincide with the top endpoint of the angled line to create a closed profile.

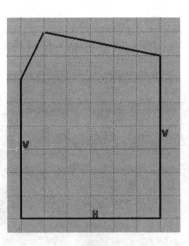

14) Perform a **Sketch Analysis** and notice that the sketch is now closed.

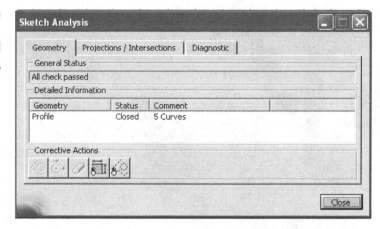

15) Draw a **Circle** inside the profile as shown in the figure.

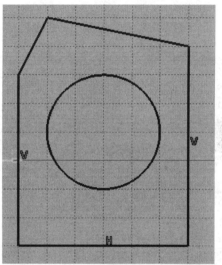

16) **Exit** the *Sketcher* and **Pocket** to the sketch to a depth of **20 mm**. Notice that the outer profile is pocketed and the inner profile is not.

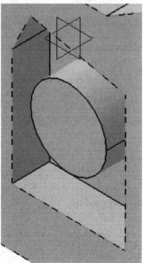

17) Enter the **Sketcher** 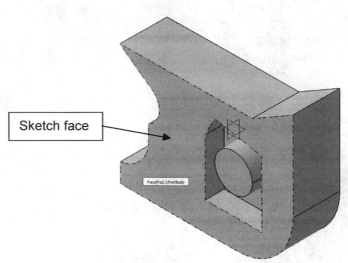 on the left face of the part.

Sketch face

18) On your own, do your best to create the **Elongated Hole** 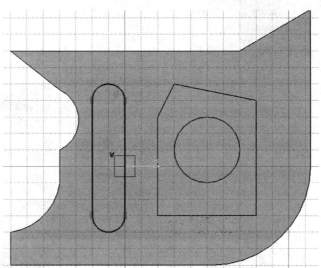 shown. Remember to read the prompt line for help. The *Elongated Hole* icon is stacked under the *Rectangle* icon.

19) **Exit** the *Sketcher* and **Pocket** to the sketch using the option of **Up to Last**.

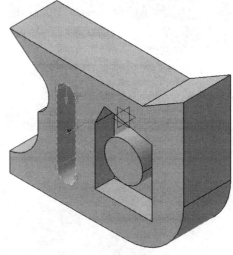

Section 3: Solving for the sketch status.

1) Rotate the part so that you can see the back face of the part. You may do this by using the middle + right mouse buttons, or by using the compass.

2) Enter the **Sketcher** 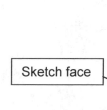 on the face indicated in the figure.

Sketch face

3) Activate your **Construction/Standard Element** 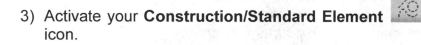 icon.

4) Use the **3D Project Element** 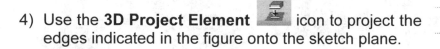 icon to project the edges indicated in the figure onto the sketch plane.

5) Deselect all.

6) Deactivate your **Construction/Standard Element** icon.

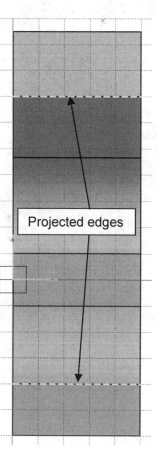

Projected edges

7) Draw the **Rectangle** 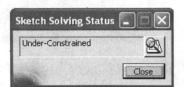 shown making sure to make coincidence constraints with the top and bottom projected lines. You will know if a coincident constraint is applied if a double circle symbol appears. Notice that the top and bottom lines are green. They are constrained.

8) Select the **Sketch Solving Status** 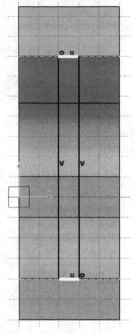 icon. A *Sketch Solving Status* window will appear and tell you that the sketch is under-constrained. **Close** the *Sketch Solving Status* window.

9) **Constrain** 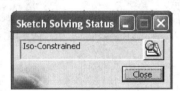 the rectangles width (*10 mm*) and distance from the side of the part (*20 mm*) as shown in the figure. Notice that once the constraints are applied, the rectangle turns green.

10) Select the **Sketch Solving Status** 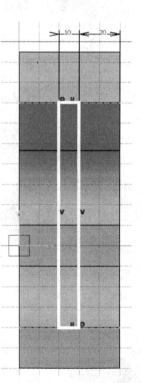 icon. A *Sketch Solving Status* window will appear and tell you that the sketch is iso-constrained (completely constrained). **Close** the *Sketch Solving Status* window.

11) **Constrain** 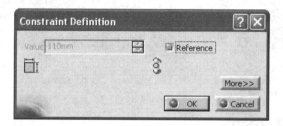 the rectangles height as shown in the figure. Notice that the dimension is purple indicating that it is a duplicate dimension.

12) Use the **Sketch Solving Status** icon to determine the sketch status.

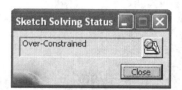

13) Double click on the duplicate height dimension. In the *Constraint Definition* window activate the **Reference** toggle. Notice the parentheses now enclose the dimension value and the rectangle again becomes iso-constrained.

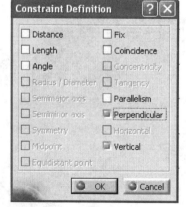

14) Apply a perpendicular constraint between the left and right vertical lines of the rectangle. Select the left and right vertical lines using the **CTRL** key. Select the **Constraint Definition** icon and activate the **Perpendicular** toggle. Notice that part of the rectangle turns pink or red indicating inconsistent constraints.

15) Use the **Sketch Solving Status** icon to determine the sketch status.

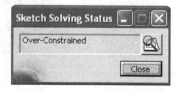

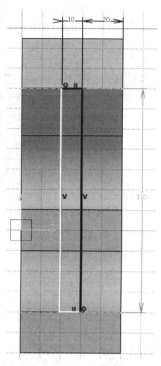

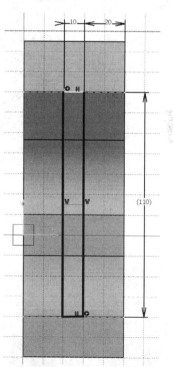

16) Select the green perpendicular constraint line between the left and right sides of the rectangle and delete it. The rectangle should now be iso-constrained.

17) **Exit** the *Sketcher* and **Pad** the rectangle up to the curved surface. In the *Pad Definition* window;
- Type: **Up to Surface**,
- Limit: select the curved surface as shown in the figure.
- If you get an error window, select the **Reverse Direction** in the *Pad Definition* window.

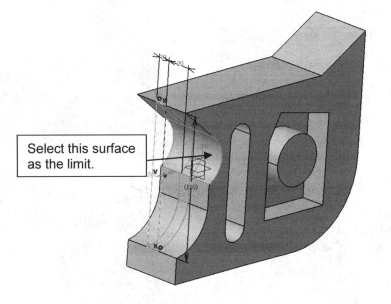

Select this surface as the limit.

Chapter 2: SKETCHER

Tutorial 2.4: Modifying Geometries & Relimitations

Featured Topics & Commands

Prerequisite Knowledge & Commands

- Entering workbenches
- Entering and exiting the *Sketcher* workbench
- Work modes (*Sketch tools* toolbar)
- Creating profiles
- Applying dimensional and geometric constraints
- Editing a preexisting sketch
- Simple pads and pockets
- Rotating the part using the mouse or compass

Modifying geometries using the mouse

The mouse can be used to move different elements of a sketch just by clicking and dragging that element. The way in which the element moves depends on the constraints that are present. For example, a horizontally constrained line can be moved anywhere but it will always remain horizontal.

The Operation toolbar

The *Operation* toolbar contains commands that allow you to modify and transform elements of a sketch. The commands contained in the *Operation* toolbar from left to right are

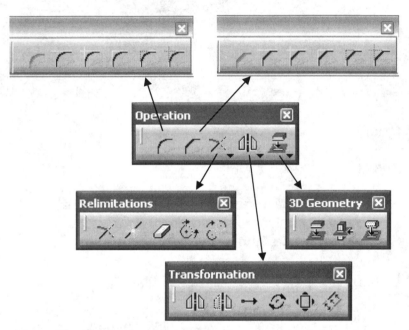

- Corner: The *Corner* or fillet command creates a rounded corner. When selected, five options appear in the *Sketch tools* toolbar. Each option is a different method of dealing with the corner after the fillet is applied. The options from left to right are: *Trim all Elements, Trim First Element, No Trim, Standard Line Trim, Construction Line Trim*, and *Construction Lines No Trim*.
- Chamfer: The *Chamfer* command creates an angled corner. When selected, five options appear in the *Sketch tools* toolbar. Each option is a different method of dealing with the corner after the chamfer is applied. The options from left to right are: *Trim all Elements, Trim First Element, No Trim, Standard Line Trim, Construction Line Trim* and *Construction Lines No Trim*.
- Relimitations toolbar: This toolbar contains commands that allow you to re-limit or modify 2D geometries. It contains commands such as *Trim, Break*, and *Close*.
- Transformation toolbar: This toolbar contains commands that allow you to duplicate and modify existing geometries. The commands from left to right are: *Mirror, Symmetry, Translate, Rotate, Scale*, and *Offset*.
- 3D Geometry toolbar: This toolbar contains commands that allow you to use 3D geometries to create 2D sketches. The commands from left to right are: *Project 3D Elements, Intersect 3D Elements*, and *Project 3D Silhouette Edges*.

The Relimitations toolbar

The *Relimitations* toolbar contains commands that allow you to re-limit or modify 2D geometries. The commands from left to right are

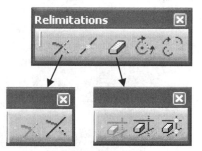

- Trim: The *Trim* command allows you to delete unwanted portions of an element. The *Trim* command allows you to dynamically choose how you want the element trimmed. When selected, two options appear in the *Sketch tools* toolbar: *Trim All Elements* and *Trim First Element*. If you are unfamiliar with this command, it may be a bit confusing.
- Break: The *Break* command allows you to break one element into two.
- Quick Trim: The *Quick Trim* command allows you to delete unwanted portions of an element. When selected, three options appear in the *Sketch tools* toolbar: *Break and Rubber In, Break and Rubber Out*, and *Break and Keep*. When the *Break and Rubber In* option is selected, the selected element will be automatically trimmed between the two closest limiting elements. When the *Break and Rubber Out* option is selected, the selected element will be kept and the elements on the outside of the two closest limiting elements will be automatically trimmed. When the *Break and Keep* option is selected, the selected element will be broken at the two closest limiting elements.
- Close: The *Close* command creates a complete circle out of an arc. This command may also be used on ellipses.
- Complement: The *Complement* command erases an arc and creates its complement. This command may also be used on ellipses.

Tutorial 2.4 Start: Part Modeled

Two different views of the part modeled in this tutorial are shown on the right. The main focus of this tutorial is to modify and relimit geometries. While constructing the first sketch, the mouse will be used to move and modify the sketch. Modifications to the main pad will be constructed using the commands located in the *Operation* and *Relimitations* toolbars.

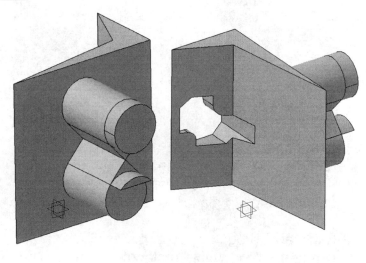

Section 1: Modifying geometries with the mouse.

1) Open a **New... Part** drawing and, if asked, name it **Angled Pin**.

2) **Save** your drawing as **T2-4.CATPart**.

3) Enter the **Sketcher** on the **xy plane**.

4) Activate **Geometrical** and **Dimensional** Constraints (orange), and deactivate the **Snap to Point** (blue).

5) Draw a **Rectangle** similar to that shown in the figure. Since the geometrical constraints are active, horizontal and vertical constraints are automatically applied.

6) Deselect all.

7) Using your mouse, click and drag the right side vertical line to the left. Notice that the horizontal and vertical constraints are maintained, but the rectangle may move up and down because it is not constrained in space.

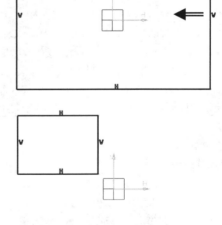

8) Click and drag the bottom right corner point up and to the left.

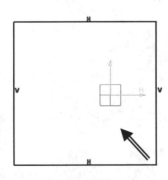

9) **Constrain** the **left** and **top** lines of the rectangle as shown. (In this instance, only select one line when constraining. Do not select 2 lines to place one dimension.)

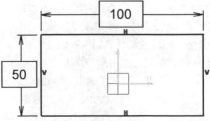

10) Click and drag the rectangle to the center. Notice that the whole rectangle moves. The rectangle is completely constrained with respect to itself, but it is not constrained in space and therefore is not green.

11)**Constrain** the rectangle in space as shown. Notice that the rectangle is now iso-constrained and is green.

12)Try to move the rectangle with the mouse. You should not be able to.

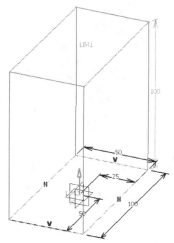

13)**Exit** the *Sketcher*

and **Pad** the sketch to a length of **100 mm**.

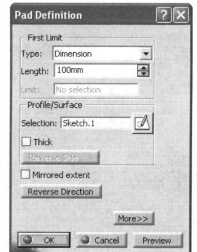

14)Edit Sketch.1.

15)Delete the 50 mm constraint that fixes the rectangle to V axis.

16)Delete the bottom horizontal (H) and right side vertical (V) constraints, and the constraints (dimensions) locating the rectangle in space.

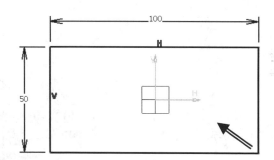

17)Click and drag the bottom right corner point up and to the left.

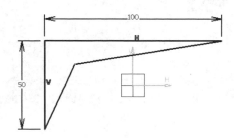

18)**Exit** ⬆ the *Sketcher.* Notice that the Pad.1 is automatically updated to reflect the changes made in the sketch.

Section 2: Re-limiting geometries

1) Deselect all.

2) Enter the **Sketcher** 🖉 on the right face of the part.

3) Pull out the *Relimitations* toolbar.

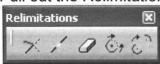

4) Draw and **Constrain** the following profile using the following commands: **Circle** ⊙ , **Bi-Tangent Line** . When completed, your sketch should be iso-constrained. You may not need to place the 50 mm constraint if you made your circles coincident with the V axis.

5) Trim the part of the circles that lie between the tangent lines. Double click on the **Quick Trim** icon and then select the parts of the circles that lie between the tangent lines as indicated in the figure. Hit the **ESC** key to exit the *Quick Trim* command.

(Note: If we tried to pad the sketch as is, CATIA would give us an error of self intersecting elements. Therefore, we need to break the tangent lines where they cross.)

6) Select the **Break** icon and then select the lines one after the other. Read the prompt line to understand why you are selecting the lines. Select the **Break** icon again. This time select the lines in reverse order.

7) **Exit** the *Sketcher* and

   Pad the sketch to a length of **40 mm**.

8) Deselect all.

9) Enter the

   Sketcher on the face of one of the padded circles as in the figure.

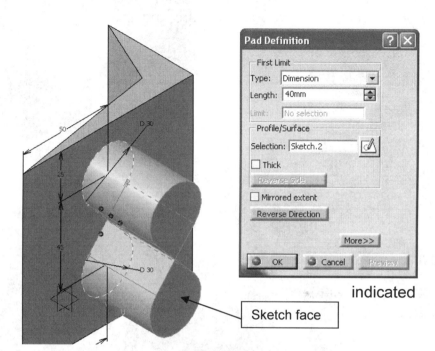

Sketch face

indicated

10) Use the **3D Project Elements** icon to project both arcs onto the sketch plane. Note that they are yellow and still associated with the part.

(Note: We are going to use the *Close* and *Complement* commands on these projected arcs. These commands can not be performed on associated (yellow) elements. Therefore, we need to disassociated or isolate the elements from the part.)

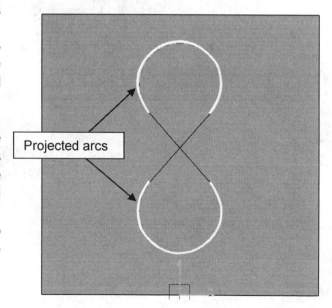

Projected arcs

11) Isolate the projected elements. Use the **CTRL** key and mouse to select both projected elements. At the top pull down menu, select **Insert** – **Operation** – **3D Geometry** – **Isolate**. Notice that the elements are no longer yellow.

12) On your own, use the **Close** command to close the top arc and the

Complement command to create a complement of the bottom arc. Then,

draw a **Line** between the endpoints of the bottom arc.

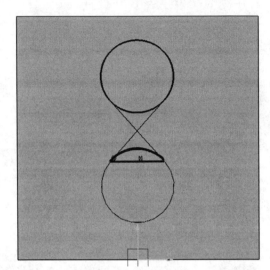

13) **Exit** the *Sketcher* and **Pad** the sketch to a length of *10 mm*.

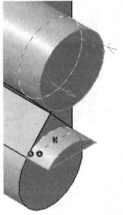

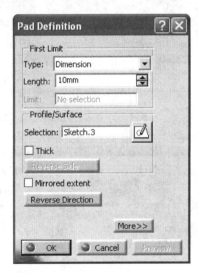

14) Save your drawing.

Section 3: Corners and Chamfers

1) Deselect all.

2) Rotate the part around and enter the **Sketcher** 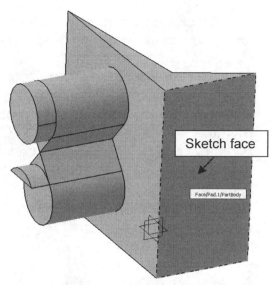 on the back face of the part as indicated in the figure.

3) Draw and **Constrain** the two **Rectangles** shown in the figure. Place the constraints on the sides shown. (Constrain the length of the rectangle sides by selecting two lines instead of one.)

4) Use **Quick Trim** to trim off the inside lines of the rectangle as shown.

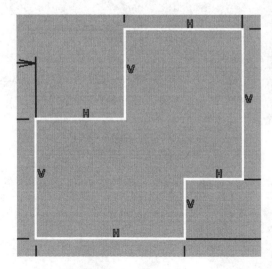

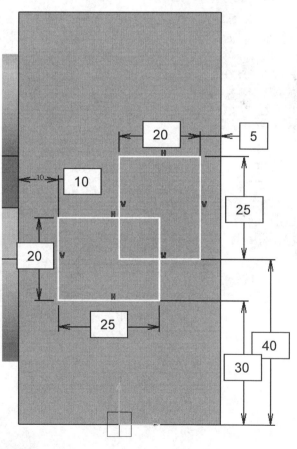

5) Apply *Corners* (fillets) to the inside corners of the sketch and change the dimension values to those shown in the figure. The corners should be applied with the *trim all* option activated. For the **8 mm** corner, select

Corner 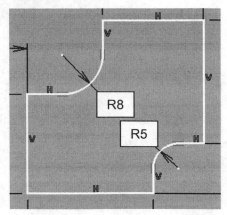 and then select the corner point. Pull the radius out and click. Change the dimension value by double clicking on the dimension. For the **5 mm** corner, select

Corner and then select both lines that make the corner.

6) Apply a **Chamfer** to the upper right corner of the sketch of length **10 mm** and angle **45°**.

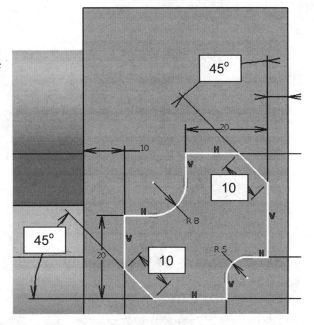

Problems? If an error window appears telling you that you are about to apply inconsistent constraints, get out of the Chamfer command and do the following. Delete the 25 mm and 20 mm height and width dimensions of the upper rectangle. Reapply the 20 mm and 25 mm **Constraints** selecting the 2 limiting lines instead of the line length.

7) Use procedures similar, re-constrain and apply a **Chamfer** to the lower left corner of the sketch of length **10 mm** and angle **45°**.

8) **Exit** 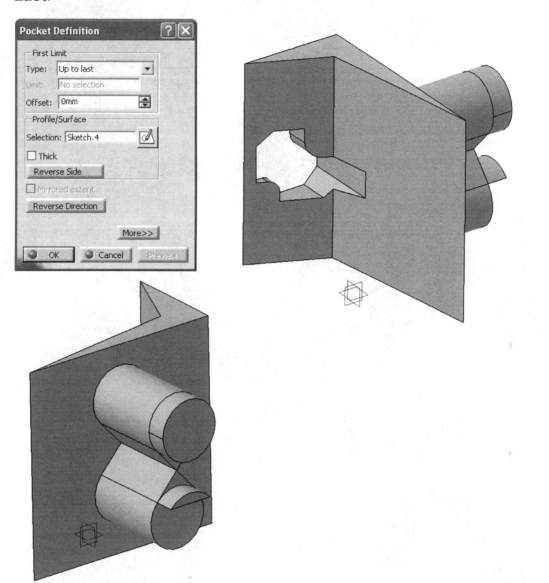 the Sketcher and **Pocket** the sketch using the option **Up to Last**.

NOTES:

Chapter 2: SKETCHER

Tutorial 2.5: Axes & Transformations

Featured Topics & Commands

Prerequisite Knowledge & Commands

- Entering workbenches.
- Entering and exiting the *Sketcher* workbench
- Work modes (*Sketch tools* toolbar)
- Creating profiles
- Applying dimensional and geometric constraints
- Editing a preexisting sketch
- Simple pads and pockets
- Rotating the part using the mouse or compass

Axes

An axis is a line of symmetry. The *Axis*

command is located in the *Profile* toolbar. When an axis line is draw it is automatically drawn as a construction element. Therefore, it does not become part of the final sketch. Axes are often used to facilitate the drawing process when using commands such as *Mirror*. It may also be used as the rotation axis when creating a *Shaft* (a revolved solid) or a *Groove* in the *Part Design* workbench.

The Transformation toolbar

Transformation tools allow you to change, or create an array of a sketch. These tools allow you to use an existing sketch to define a new sketch that is a mirror image, a rotated version, or a different size from the original sketch. Many of the transformation commands may be used in 'duplicate' mode which enables arrays to be created. The *Transformation* toolbar in located in the *Operation* toolbar. Reading from left to right, the *Transformation* toolbar contains the following commands

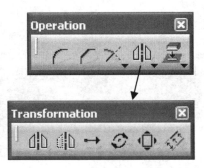

- Mirror: The *Mirror* command creates a mirror image of a sketch about a selected axis.
- Symmetry: The *Symmetry* command replaces a sketch with its mirror image.
- Translate: The *Translate* command moves a sketch from one point to another. This command may be used in duplicate mode to create a linear array.
- Rotate: The *Rotate* command rotates a sketch around a selected center point. This command may be used in duplicate mode to create a polar array.
- Scale: The *Scale* command scales a sketch relative to a selected point. This command may be used in duplicate mode which creates several instances of the sketch.
- Offset: The *Offset* command creates a sketch that is offset from the original by a specified distance. This command scales and adjusts the sketch as appropriate. Given the complex nature of this command, it does not always work as intended.

Tutorial 2.5 Start: Part Modeled

The part modeled in this tutorial is shown on the right. The outside hub is created using an *Axis* and the *Shaft* command. The spokes are created using the *Rotate* command, and the triangular *Pockets* are created using the *Mirror* command.

Section 1: Using Axes

1) Open a **New… Part** drawing and, if asked, name it ***Wheel***.

2) **Save** your drawing as ***T2-5.CATPart***.

3) Enter the **Sketcher** 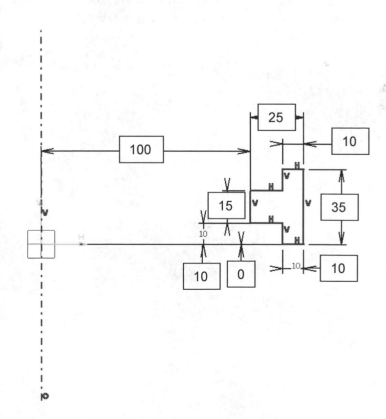 on the **yz plane**.

4) Draw an **Axis** that is on and aligned with the vertical (V) axis.

5) Draw and **Constrain** the **Profile** shown. Depending on how many geometric constraints were applied, you may or may not have to include all of the dimensional constraints shown. The sketch should be iso-constrained when complete.

6) **Exit**  the Sketcher and **Shaft** the sketch using the Sketch Axis as the rotation axis. In the *Shaft Definition* window fill in the following fields;

- First angle: **360**
- Second angle: **0**
- Profile Selection: **Sketch.1**
- Axis Selection: **Sketch Axis**

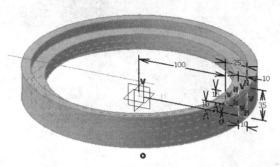

7) In the *Part Design* workbench apply fillets to the inside corners of the hub. Select

the **Edge Fillet** icon. In the *Edge Fillet Definition* window, set the fillet radius to **2 mm**. Activate the *object(s) to fillet:* field and select all 6 inside edges and then select **OK**. (UNIX users will need to hold the CTRL key down when selecting the edges.)

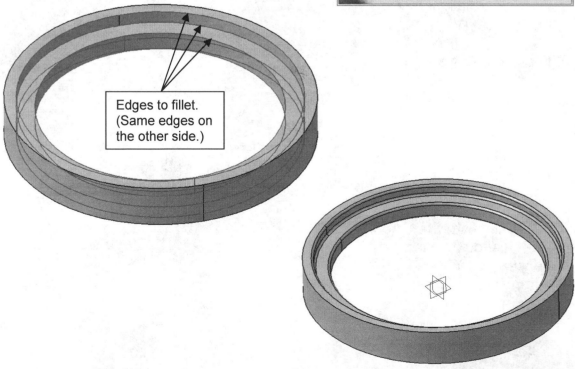

Edges to fillet.
(Same edges on
the other side.)

8) Deselect all and enter the **Sketcher** on the **yz plane**.

9) Select the **Cut Part by Sketch Plane** icon located in the *Visualization* toolbar usually in the bottom toolbar area.

10) Draw an **Axis** that is on and aligned with the V axis.

11) Draw and **Constrain** the **Profile** shown.

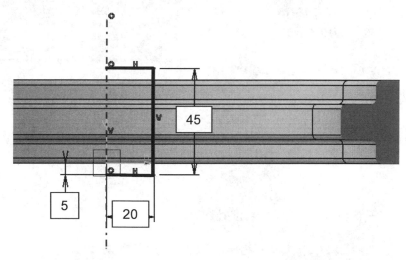

12) **Exit** the Sketcher and **Shaft** the sketch *360* degrees using the Sketch Axis as the rotation axis.

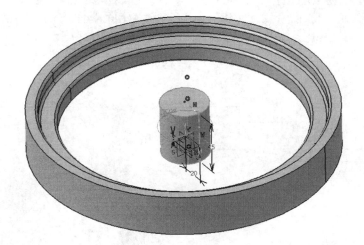

13) **Save** your drawing.

Section 2: Transformations

1) Activate the *Reference Element (Extended)* toolbar. At the top pull down menu select **View – Toolbars – Reference Elements (Extended)**. If it is not apparent where this toolbar ended up, look in the corners for any hidden toolbars.

2) Create a new sketch plane by offsetting the xy plane by 12 mm. Select the **Plane** 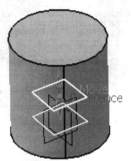 icon. In the *Plane Definition* window fill in the following fields;
   - Plane type: **Offset from plane**
   - Reference: right click and select **xy plane**
   - Offset: *12 mm*.

   Then select **OK**. A new plane should be created 12 mm from the xy plane.

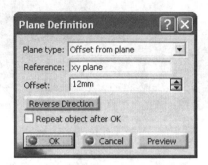

3) Enter the **Sketcher**  on the newly created plane.

4) Pull out the *Transformation* subtoolbar.

5) Draw and **Constrain** the following **Ellipse**. To constrain the major axis just click on the ellipse. To constrain the minor axis click on the ellipse, when the major axis dimension appears, right click and select **Semiminor axis**.

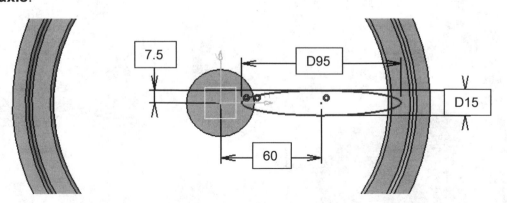

6) Array the ellipse to create a total of 8 ellipses each 45 degrees apart. Select the ellipse and then select the

Rotate 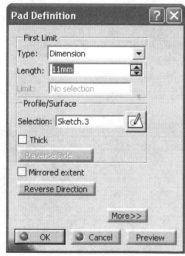 icon. In the *Rotation Definition* window select **7** Instance(s), and activate the **Duplicate mode**. Notice that the Angle field does not allow you to enter a value yet. Look at the Prompt line. Select the center point to be the coordinate origin, enter an angle of ***45deg*** and then select **OK**. (UNIX users need to hold the CTRL key down when the center point is selected.)

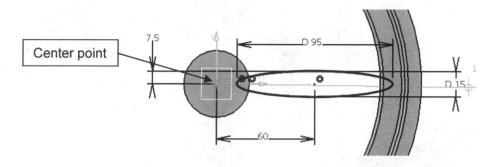

7) **Exit** 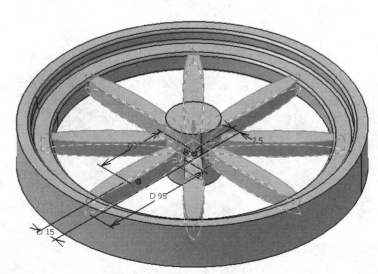 the Sketcher and **Pad** the sketch to a length of ***11 mm***.

8) **Save** your drawing.

9) Select the **Edge Fillet** 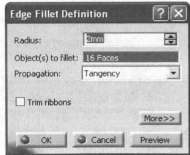 icon. In the *Edge Fillet Definition* window, set the fillet radius to *2 mm*. With the object(s) to fillet: field active select all 16 side faces of the padded ellipsis and then select **OK**. (UNIX users will need to hold the CTRL key down when selecting the faces.)

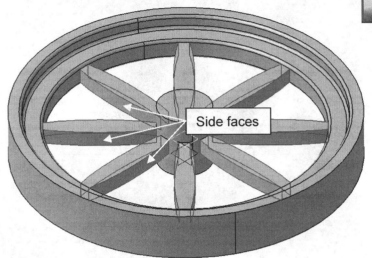

10) Enter the **Sketcher** on the top face of the center hub.

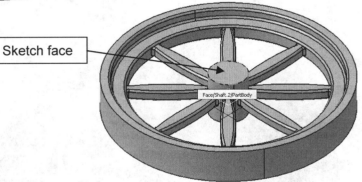

11) Draw and **Constrain** the following **Axis** and **Profile**.

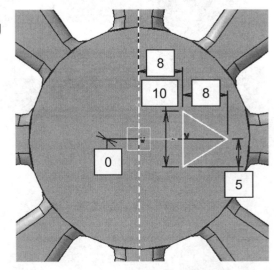

12) Use the *Symmetry* command to mirror the original sketch to the other side of the axis. Select all elements of the triangle, select the **Symmetry** 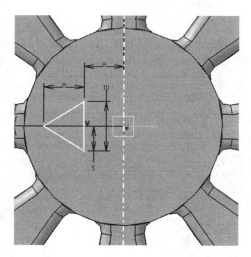 icon and then select the axis.

13) On your own, use the **Mirror** 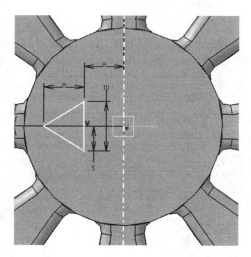 command to create a copy of the sketch on the other side of the axis.

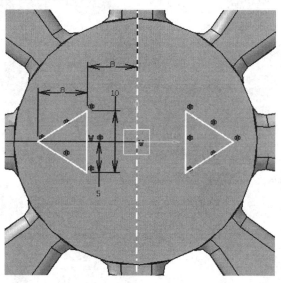

14) **Rotate** both triangles through a distance of *90deg* creating *1* instance in the **Duplicate mode**.

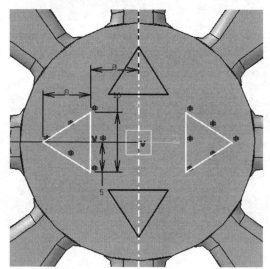

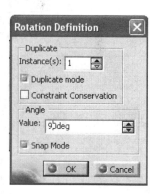

15) **Exit**  the Sketcher and **Pocket** the sketch using the option **Up to last**.

(Problems? If your sketch disappears, *Copy* and *Paste* the Sketch as described in the preface.)

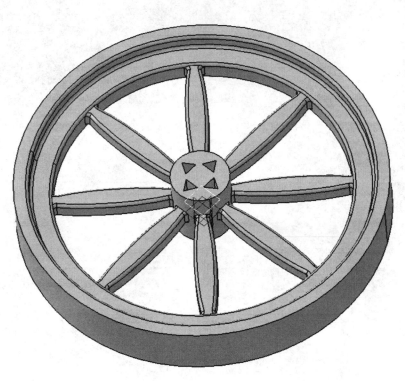

Chapter 2: SKETCHER

Tutorial 2.6: Operations on 3D Geometries & Sketch planes

<u>Featured Topics & Commands</u>

<u>Prerequisite Knowledge & Commands</u>

- Entering workbenches
- Entering and exiting the *Sketcher* workbench
- Work modes (*Sketch tools* toolbar)
- Creating profiles
- Applying dimensional and geometric constraints
- Editing a preexisting sketch
- Simple pads and pockets
- Rotating the part using the mouse or compass

The 3D Geometry toolbar

Many times it is necessary to relate the current sketch to the existing solid part. There are two ways of doing this. You can constrain the sketch using the edges of the part as a point of reference, or you can project the part's edges and silhouettes onto the sketch plane. This second method of projection has many advantages. It is quick (you don't have to redraw the sketch) and the projection is associative. An associative projected sketch is connected to the existing part and automatically updates when the part is changed. The default color of an associative sketch is yellow. The *3D Geometry* toolbar is located in the *Operation* toolbar. The commands in the *3D Geometry* toolbar from left to right are

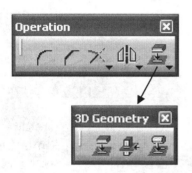

- <u>Project 3D Elements:</u> This command allows you to project edges of the solid part onto the sketch plane.
- <u>Intersect 3D Elements:</u> This command allows you to project the intersection between the part and the sketch plane.
- <u>Project 3D Silhouette Edges:</u> This command allows you to create a silhouette of the solid part. You can only create a silhouette edge projection from a surface of revolution whose axis is parallel to the sketch plane.
- <u>Isolate:</u> *Isolate* is a command that is not located in the *3D Geometry* toolbar but is used in conjunction with the above three commands. *Isolate* breaks the link between the projected sketch and the part. Once the *Isolate* command has been applied to a sketch, the sketch will no longer be associative. To access this command select *Insert – Operation – 3D Geometry – Isolate*.

The Reference Element toolbar

Thus far our sketch planes have mainly been the 3 principle planes or a face of the part. Many times we need a sketch plane that can not be defined by the original coordinate system or the part. The commands located in the *Reference Element* toolbar allows us to create planes and other reference elements that may be used as the sketch plane or help in the creation of a sketch plane. These elements are not standard elements and will not become part of your solid. The *Reference Element* toolbar is located in the *Part Design* workbench and is not normally active. To activate this toolbar select *View – Toolbars – Reference Elements (Extended)*. The commands located in the *Reference Element* toolbar from left to right are

- <u>Point:</u> Allows you to create a reference point.
- <u>Line:</u> Allows you to create a reference line. Mainly used to define a direction for another command.
- <u>Plane:</u> Used to create a reference plane to be used as a sketch plane. There are many ways to create a reference plane. A couple examples include; a plane offset from an existing plane or part face, and a plane at an angle to another plane of part face.

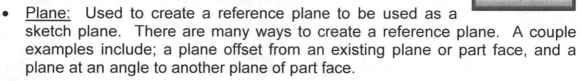

Tutorial 2.6 Start: Part Modeled

The part modeled in this tutorial is shown on the right. This part is constructed using projections and reference planes. Notice that several reference planes have been created to facilitate the drawing process.

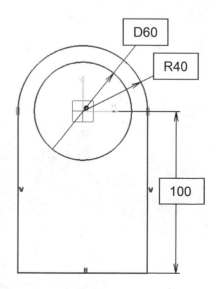

Section 1: 3D Project Elements

1) Open a <u>**New...**</u> **Part** drawing and, if asked, name it ***Angled Channel***.

2) **Save** your drawing as ***T2-6.CATPart***.

3) Enter the **Sketcher** 🖋 on the **yz plane**.

4) Draw and **Constrain** ▣ the following **Profile** 🔓 and **Circle** ⊙ shown. Notice that the circle's center is coincident with the origin.

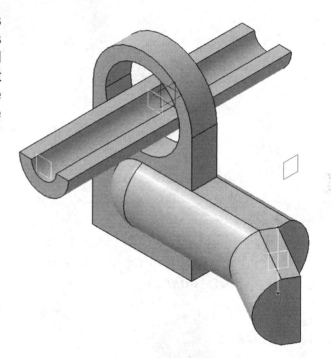

5) **Exit** 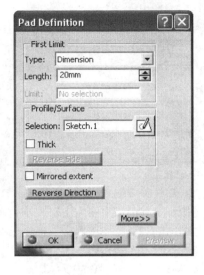 the *Sketcher* and **Pad** the sketch to a length of *20 mm*.

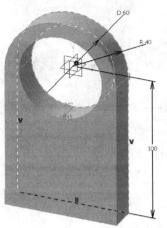

6) Create a reference plane that is offset from the yz plane by 100 mm. If the *Reference Element (Extended)* toolbar is not active, activate it. At the top pull down menu select **View – Toolbars – Reference Elements (Extended)**.

Select the **Plane** icon. In the *Plane Definition* window fill in the following fields;

- `Plane type`: **Offset from plane**
- `Reference`: right click and select **yz plane**
- `Offset`: *100 mm*.

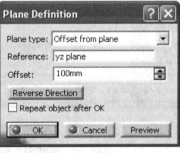

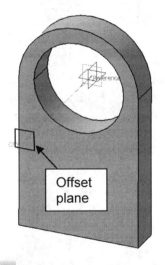

Offset plane

7) Enter on the **Sketcher** on the newly created plane.

8) Select the **Isometric View** icon.

9) Display the **Grid** .

10) Use the **3D Project Elements** to project the outline of the hole onto the sketch plane. Notice that the projection is yellow. This means that it is associated with the part.

11) Offset the projection 10 mm to the inside. Select the **Offset** icon (located in the *Transformation* toolbar), select the projection, move the mouse to the inside of the circle and click, and then double click on the dimension and change its value to **10 mm**.

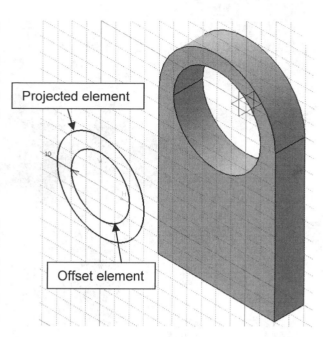

Projected element

Offset element

12) Draw a horizontal **Line** between quadrants of the outside projected circle.

13) Use the **Quick Trim** command to trim off the top half of the two circles as well as the portion of the line inside the smaller circle. The trimmed portions of the associated elements will turn into construction elements.

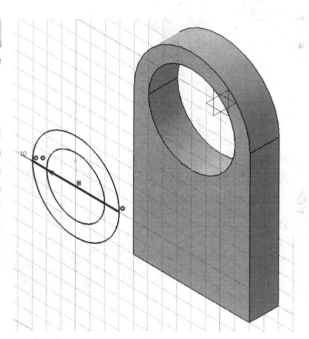

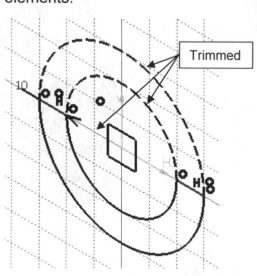

Trimmed

14) **Exit** 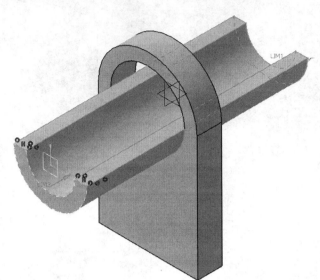 the *Sketcher* and **Pad** the sketch to a length of **180 mm** in the **Reverse Direction**.

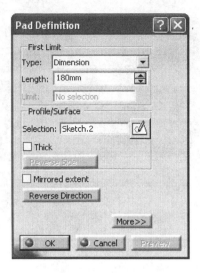

15) **Save** your drawing.

16) Edit Sketch.1.

17) Change the hole diameter to **40 mm**.

18) **Exit** the *Sketcher.* Notice that Pad.2 which is based on the projection updates automatically.

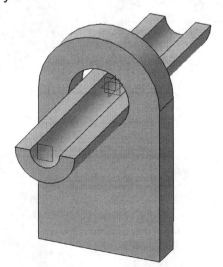

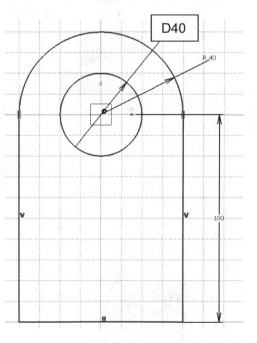

19) Edit Sketch.2.

20) Use a selection window to select all the elements in Sketch.2. Then at the top pull down menu, select **Insert – Operation – 3D Geometry – Isolate**. Notice that the projections are no longer yellow. They are not associated with the part anymore.

21) **Exit** the *Sketcher* and edit Sketch.1.

22) Change the Sketch.1 hole diameter back to **60 mm**.

23) **Exit** the *Sketcher*. Notice that Pad.2 does not update this time.

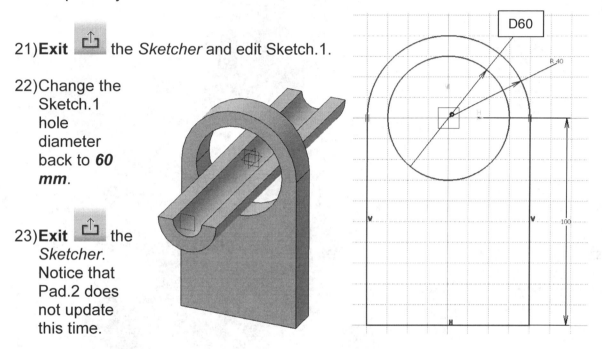

Section 2: Project 3D Silhouette Edges

1) Offset a **Plane** **10 mm** from the **yz plane**.

2) Enter on the **Sketcher** on the newly created plane.

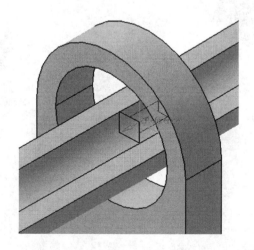

3) Draw and constrain the following Sketch.

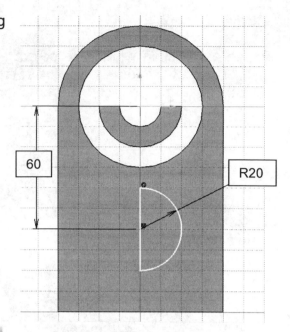

4) **Exit** the *Sketcher* and **Shaft** the sketch through **360deg** using the **V Direction** as the rotation axis.

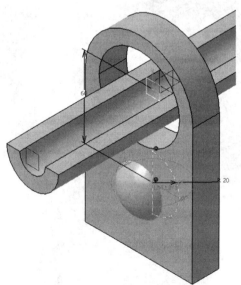

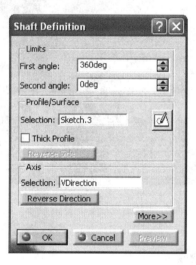

5) Offset a **Plane** 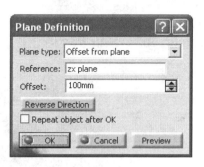 *100 mm* from the **zx plane**.

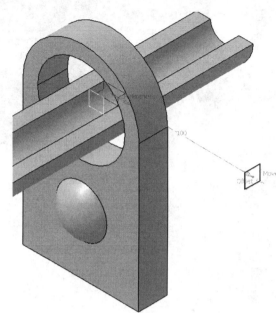

6) Enter on the **Sketcher** on the newly created plane.

7) Select the **Isometric View** icon.

8) Try using **3D Project Elements** to project the Shafted sphere. You can't do it. All you get is a line. Delete the line that was created.

9) Use **Project 3D Silhouette Edges** to project the silhouette of the two protruding parts of the sphere.

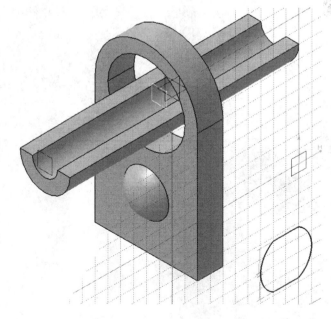

10) Draw **Lines** to connect the top ends and bottoms ends of the projected arcs.

11) Delete the parallel constraint between the two lines if one appears.

12) **Exit** 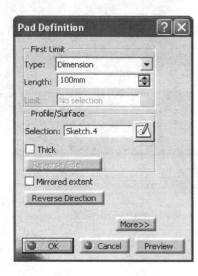 the *Sketcher* and **Pad** the sketch to a length of **100 mm** in the **Reverse Direction**.

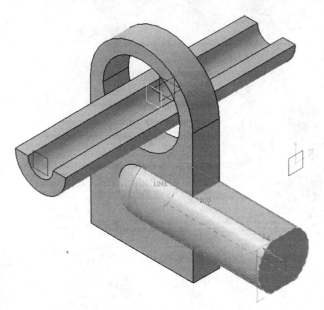

Pad Definition

First Limit

Type: Dimension

Length: 100mm

Limit: No selection

Profile/Surface

Selection: Sketch.4

☐ Thick

Reverse Side

☐ Mirrored extent

Reverse Direction

More>>

OK   Cancel   Preview

Problems? If CATIA will not allow you to Pad your sketch, it may have created a point in the middle of the sketch. Edit the sketch and turn the point into a Construction Element.

13) Edit Sketch.3. This is the sketch associated with the *Shaft*.

14) Change the radius dimension to **25 mm**.

15) **Exit** the *Sketcher*. Notice that Pad.3 automatically updates.

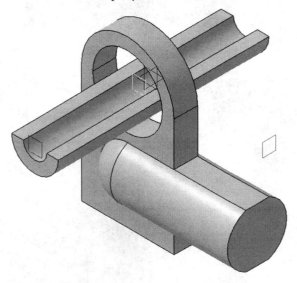

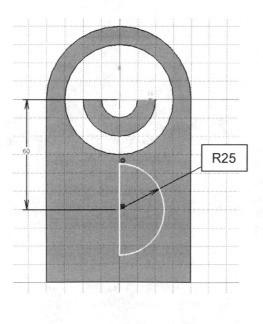

R25

Section 3: Intersect 3D Elements.

1) Create two reference points. Select the **Point**

   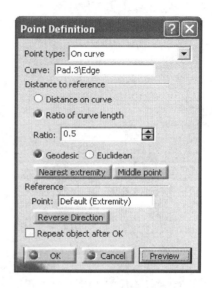

   icon. In the *Point Definition* window fill in the following fields;
   - `Point type:` **On curve**
   - `Curve:` select the top line shown in the figure
   - `Distance to reference:` activate **Ratio of curve length**
   - `Ratio:` *0.5*.
   - `Point:` Clear the selection

   Create a second point on the bottom line in a similar fashion.

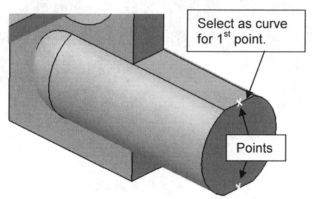

Select as curve for 1st point.

Points

2) Create a reference **Line**

   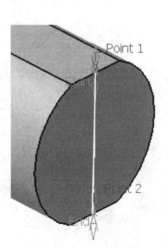

   between the two points.

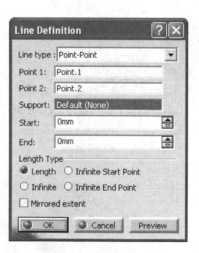

3) Select the **Plane** 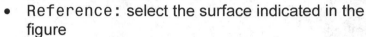 icon. In the *Plane Definition* window fill in the following fields;

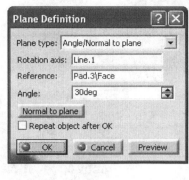
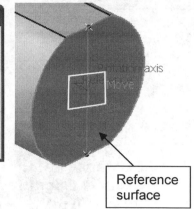

Reference surface

- **Plane type: Angle/Normal to plane**
- **Rotation axis:** click on the **Line.1** (the reference line)
- **Reference:** select the surface indicated in the figure
- **Angle: 30deg**.

4) Enter on the **Sketcher** 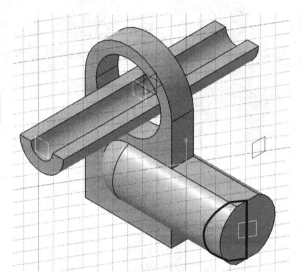 on the newly created plane.

5) Select the **Isometric View** icon.

6) Select the **Intersect 3D Elements** icon. Then select Pad.3 (the last pad) in the specification tree.

16) **Exit** the *Sketcher* and **Pad** the sketch to a length of **50 mm** in the **Reverse Direction**.

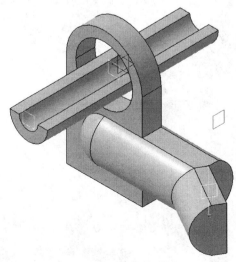

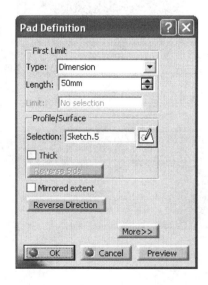

Chapter 2: SKETCHER

Tutorial 2.7: Points & Splines

Featured Topics & Commands

Prerequisite Knowledge & Commands

- Entering workbenches
- Entering and exiting the *Sketcher* workbench
- Work modes (*Sketch tools* toolbar)
- Creating profiles
- Applying dimensional and geometric constraints
- Editing a preexisting sketch
- Simple pads and pockets
- Rotating the part using the mouse or compass

Points

Points are used to define some aspects of 2D geometries. For instance, a line is defined by two end points, and a circle is defined by a center point and a radius. When creating geometries such as lines, circles, rectangles and profiles, points are automatically created by CATIA. However, points may also be created by the user. User created points are used when creating splines and wireframes. The *Point* sub-toolbar is located within the *Profile* toolbar. Reading from left to right, the *Point* sub-toolbar contains the following commands

- Point by Clicking: Allows you to create a point by clicking anywhere on the screen. The position of this point may be changed later by double clicking on and entering a new position in the *Point Definition* window.
- Point by Using Coordinates: Allows you to create a point by entering a Cartesian or polar coordinate in the *Point Definition* window.
- Equidistant Points: Allows you to create a specified number of equidistant points along a previously drawn curve.
- Intersection Point: Allows you to create a point at the intersection of two elements.
- Projection Point: Projects (in 2D) a point onto a pre-existing profile. The projection line which connects the point with the profile creates a 90° angle with the profile.

Splines

Splines are used when a profile is curved and complex. Such profiles are too complex and are difficult to create using arcs and conics. Reading from left to right, the *Spline* toolbar contains the following commands.

- Spline: A spline is a curved profile defined by three or more points. The tangency and curvature radius at each point may be specified.
- Connect: The connect command connects two points or profiles with a spline.

Tutorial 2.7 Start: Part Modeled

The part modeled in this tutorial is a cam. The cam profile is created by using construction points and a spline. The cam profile is then modified by entering the *Spline Definition* window and adding and modifying the points used to create the spline. The cam is shown at two different stages of the tutorial. The original cam profile is shown below. The cam profile graph shows the follower's displacement versus the rotation angle. The cam has a base circle radius of 1.5 inches.

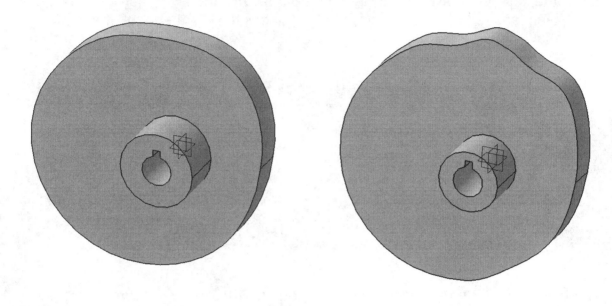

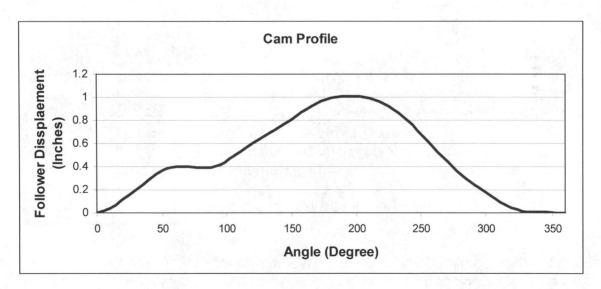

Section 1: Entering points.

1) Open a **New... Part** drawing, name it *Cam* and enter the **Sketcher** 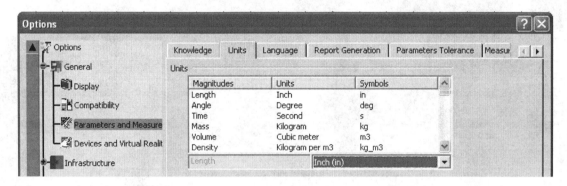 on the **yz plane**.

2) Save your drawing as *T2-7.CATPart*.

3) At the top pull down menu, select **Tools – Options...** Deactivate the grid display and snap to point options. In the left side menu of the *Options* window, select **Parameters and Measure**. Click on the **Units** tab and set the length units to **Inches**.

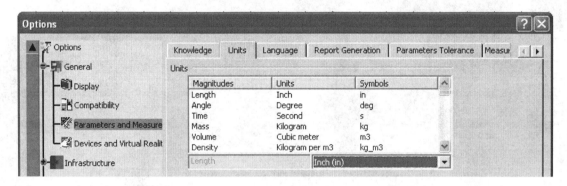

4) Activate the **Construction/Standard Element** icon. We want the points that we will be creating to be construction elements.

5) Double click* on the **Point by Using Coordinate** icon. In the *Point Definition* window click on the **Polar** tab and enter a radius of *1.5in* and an Angle of *0deg*. Then select **OK**. When the next *Point Definition* window appears, click on the **Polar** tab and enter a radius of *1.7in* and an Angle of *30deg*. Continue filling in the *Point Definition* window with the following Cam profile data. Remember to add the base circle radius to each displacement point. We have already entered the first two points. Start with the third.

---

* Double clicking allows you to continually enter points that are relative to the origin. If you don't double click you will end up entering points that are relative to the last point created unless you deselect the last point each time.

Angle (Degrees)	Follower Displacement (Inches)
0	0
30	0.2
60	0.4
90	0.4
105	0.49
120	0.6
135	0.705
150	0.81
165	0.92
180	1
210	1
225	0.92
240	0.79
255	0.62
270	0.45
285	0.3
300	0.18
315	0.07
330	0.01

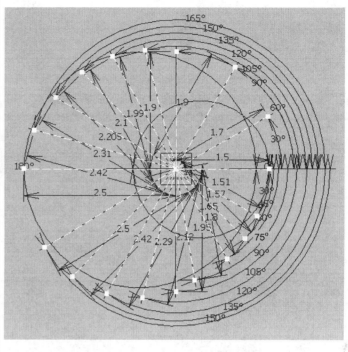

Section 2: Creating and editing splines.

1) Deselect all and then deactivate the **Construction/Standard Element** icon.

2) Click on the **Spline** icon. Move your mouse to the first created point (1.5in<0deg). When a bull's-eye symbol appears click. Select each successive point until you return to the first point. Double click on the first point to end the spline.

Bull's-eye symbol

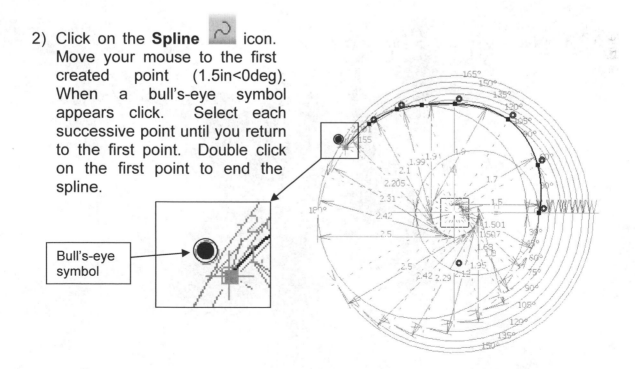

3) **Exit** 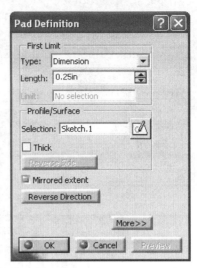 the *Sketcher* and **Pad** the sketch to a length of *.25in* in both directions (**Mirrored extent**).

(Problems? If CATIA does not allow you to pad the spline, one of the points used to create the spline may be a standard element instead of a construction element.)

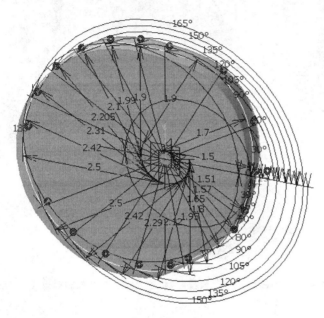

4) Enter the **Sketcher** on the **yz plane**.

5) Draw a **Circle** that is centered at the origin and has a diameter of *1.25 in*.

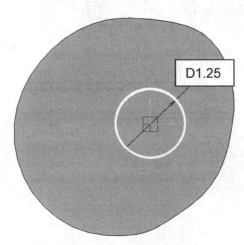

6) **Exit** the *Sketcher* and **Pad** the sketch to a length of *1in* in both directions (**Mirrored extent**).

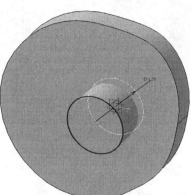

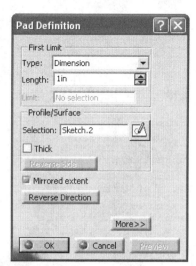

7) Enter the **Sketcher** on the front face of the center hub.

8) Draw and constrain the following sketch.

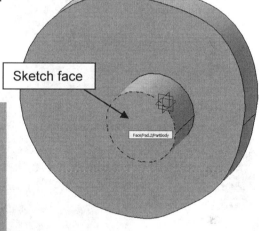

Sketch face

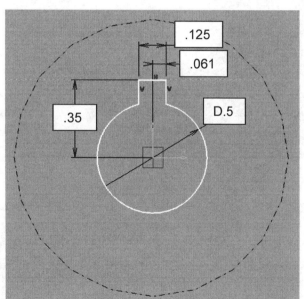

.125

.061

.35

D.5

9) **Exit** 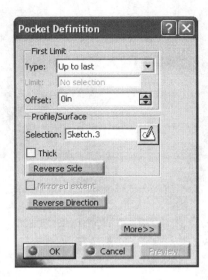 the *Sketcher* and **Pocket** the sketch using the option **Up to last**.

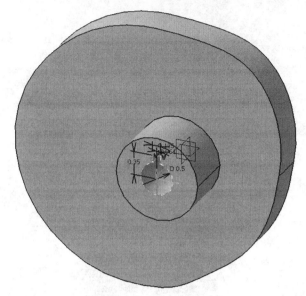

10) Save your drawing.

11) Edit Sketch.1.

12) Double click anywhere on the spline. In the *Spline Definition* window, click on **CtrPoint.4** and activate **Replace Point**. Click on the graphics screen somewhere above CtrPoint.4 making sure not to create any constraints. Click **OK**.

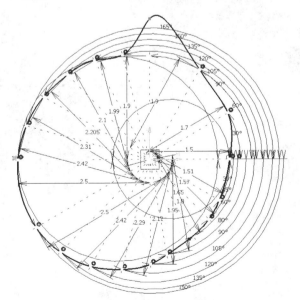

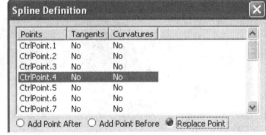

13) Double click on the newly created point. In the *Point Definition* window, click on the **Polar** tab and enter a radius of **2.1in** and an angle of **90deg**.

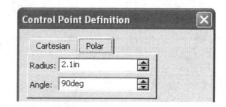

14) Double click on the spline. In the *Spline Definition* window click on **CtrlPoint.3** and activate **Add Point After**. Click somewhere between CtrlPoint.3 and CtrlPoint.4 making sure not to create any constraints.

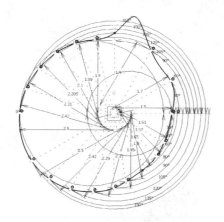 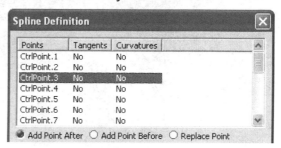

15) Edit the new point's position and enter R = **2in** and A = **80deg**.

16) Double click on the spline. In the *Spline Definition* window click on **CtrlPoint.21** the point between CtrlPoint.3 and CtrlPoint.4. Activate both the **Tangency** and **Curvature Radius** toggles. Enter a Curvature Radius of **.4in**. Look at the spline while you click **OK** to see the effect that changing the curvature radius has on the spline.

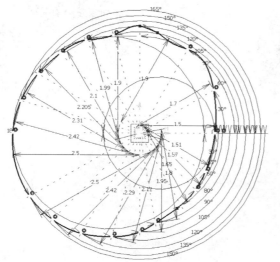 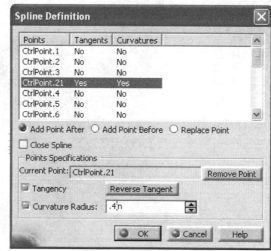

17)**Exit** 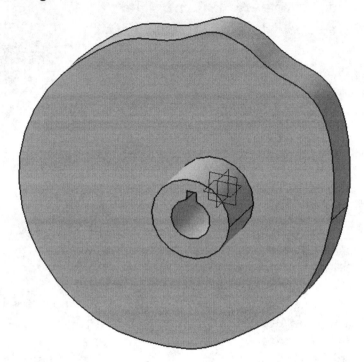 the *Sketcher.*

18)Change the diameter of inner hub from 1.25 inches to *1 inch*.

19)Save your drawing.

Chapter 2: EXERCISES

Exercise 2.1: This exercise can be performed after completing tutorial 2.1. Model the following parts using the *Snap to Point* command to estimate distances. You will be creating a 3-D models based on the 2-D orthographic projections shown.

a)   b)   c)   d)

e)   f)

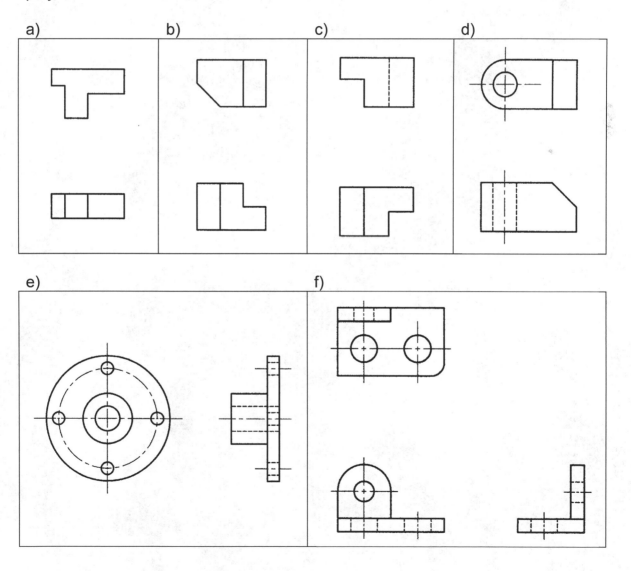

Exercise 2.2: This exercise can be performed after completing tutorial 2.2. Model the following parts applying the appropriate constraints. To set the appropriate units, select _Tools – Options... – Parameters and Measure_.

a)

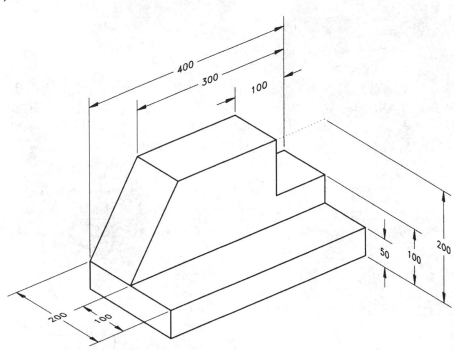

b)

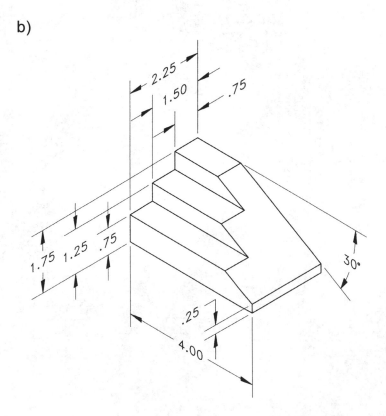

Exercise 2.3: This exercise can be performed after completing tutorial 2.3. Model the following parts applying the appropriate constraints. To set the appropriate units, select _Tools_ – _Options..._ – _Parameters and Measure_.

a)

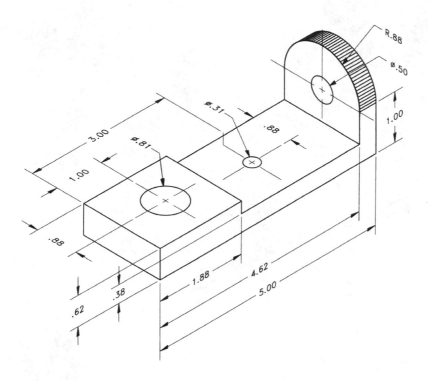

b)

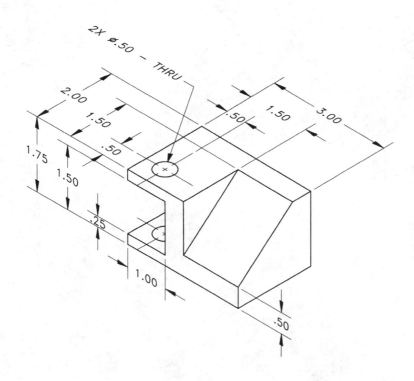

c)

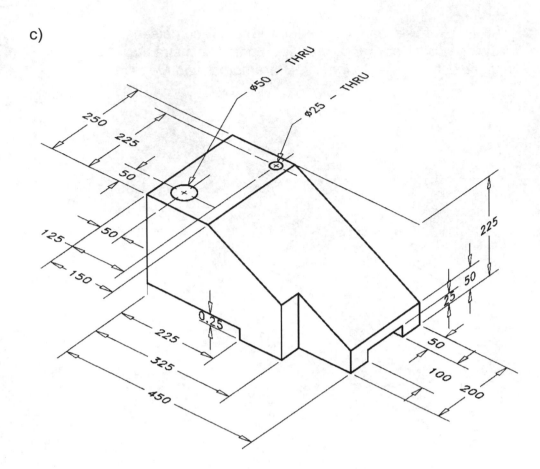

d)

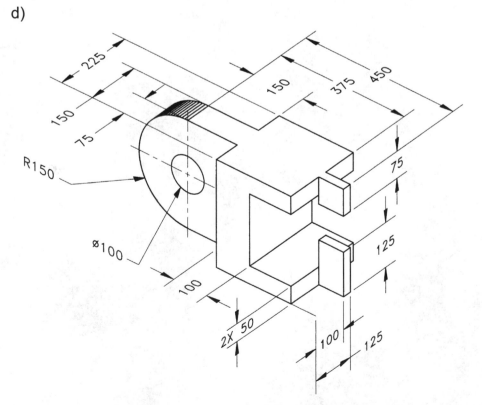

Exercise 2.4: This exercise can be performed after completing tutorial 2.4. Sketch the following parts applying the appropriate constraints. To set the appropriate units, select _Tools – Options... – Parameters and Measure._

a)

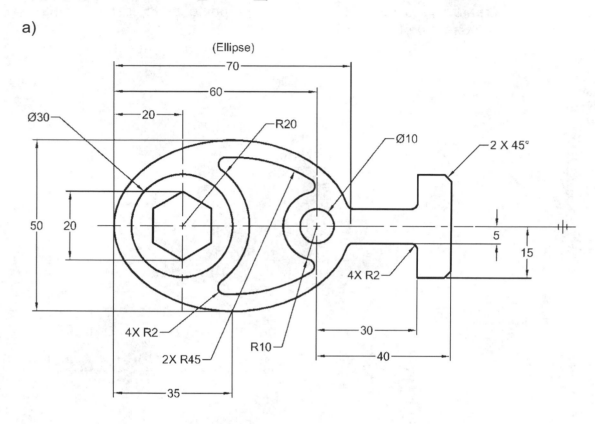

b)

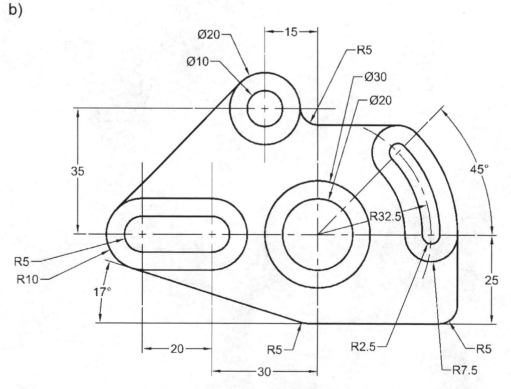

Exercise 2.5: This exercise can be performed after completing tutorial 2.5. Model the following parts applying the appropriate constraints, and transformations during the sketch phase. Use at least one transformation command to create the object. To set the appropriate units, select *Tools – Options... – Parameters and Measure.*

a)

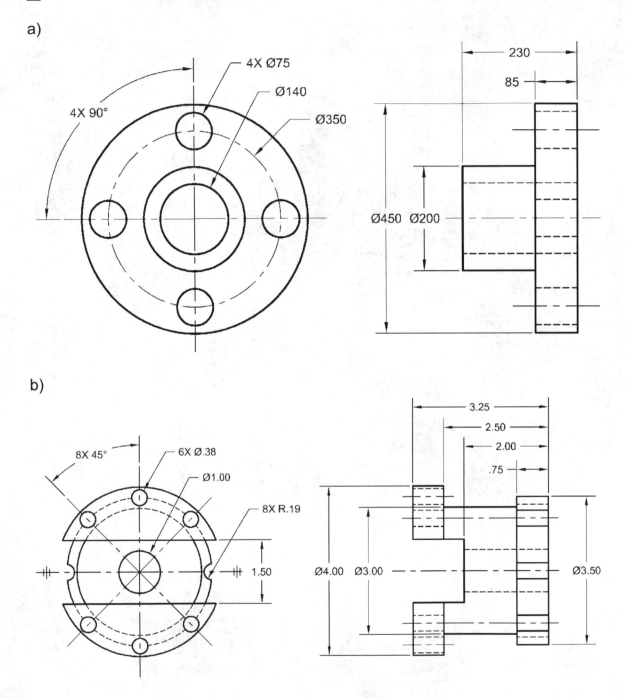

b)

Exercise 2.6: This exercise can be performed after completing tutorial 2.6. Model the following parts applying the appropriate constraints, and using reference elements and projections where necessary. To set the appropriate units, select *Tools – Options... – Parameters and Measure*.

a)

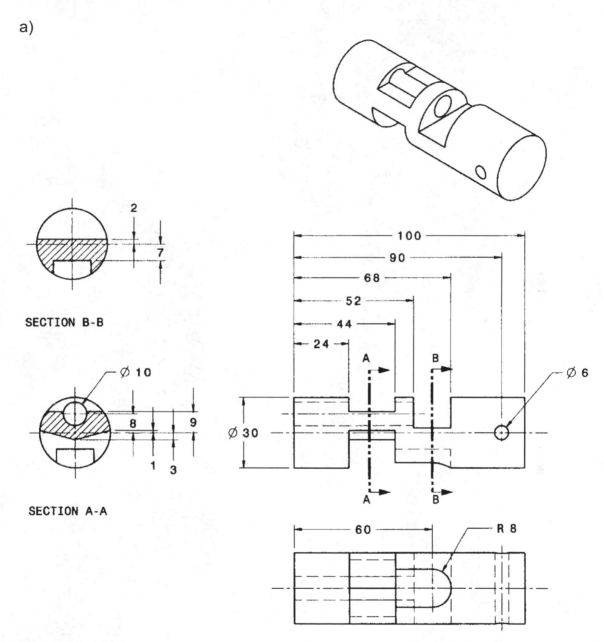

SECTION B-B

SECTION A-A

b)

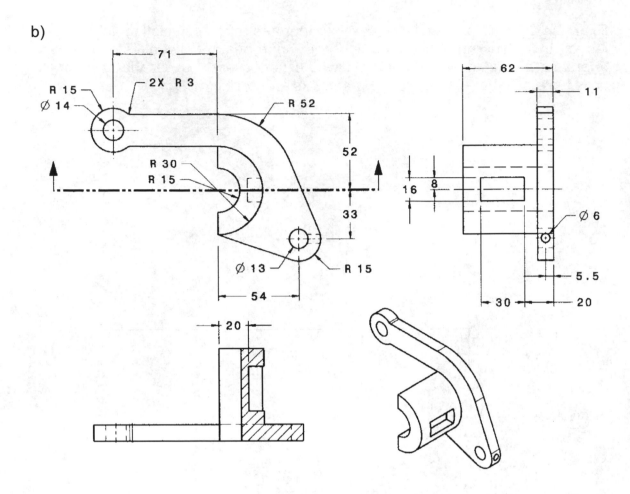

Chapter 3: PART DESIGN

Introduction

Chapter 3 focuses on CATIA's *Part Design* workbench. The reader will learn how to use sketches to create solids and how to create solids without the aid of sketches.

Tutorials Contained in Chapter 3

- Tutorial 3.1: Sketch-Based Features: Pads
- Tutorial 3.2: Sketch-Based Features: Pockets, Holes & Slots
- Tutorial 3.3: Sketch-Based Features: Shafts & Grooves
- Tutorial 3.4: Sketch-Based Features: Advanced commands
- Tutorial 3.5: Dress-Up Features: Drafts, Fillets & Chamfers
- Tutorial 3.6: Dress-Up Features: Shell, Thickness & Threads
- Tutorial 3.7: Transformation Features
- Tutorial 3.8: Boolean Operations
- Tutorial 3.9: Material Properties

NOTES:

NOTES:

Chapter 3:
PART DESIGN

Tutorial 3.1: Sketch-Based Features: Pads

Featured Topics & Commands

Prerequisite Knowledge & Commands

- The *Sketcher* workbench and associated commands
- Editing a preexisting sketch
- Simple pads and pockets
- Rotating the part using the mouse or compass

The Part Design workbench

The *Part Design* workbench is used to create and modify solids. A solid is first created using an initial sketch. Once the main solid is created, it may be modified using sketch based feature commands or using commands that do not require a sketch. The general design process is to; sketch the profile of the main pad, generate the main pad, create additional sketch based features, add dress-up features, modify features as desired, and insert new bodies for more complex parts.

The Sketch-Based Features toolbar

The commands located in the *Sketch-Based Features* toolbar use sketches to create or modify a solid. Therefore, to modify the shape of the feature, the sketch must first be modified. The depth, length of travel, rotation angle, etc... may be modified in the respective definition windows. It is possible to modify a feature (Pad, Pocket, etc...) after it has been created. Just double click on the respective feature in the specification tree and the *Definition* window will appear allowing you to change the parameters. Reading left to right, the commands in the *Sketch-Based Features* toolbar are

- Pad: *Pads* are used to add or create solid material in a linear direction.

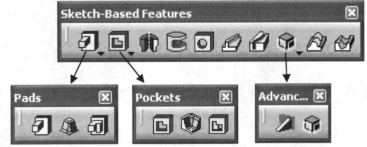

- Pocket: *Pockets* are used to remove material in a linear direction.
- Shaft: *Shafts* are used to add or create material by revolving a sketch about an axis.
- Groove: *Grooves* are used to remove material by revolving a sketch about an axis.
- Hole: The *Hole* command is used to create holes of different configurations. For example, this command could be used to create a simple blind hole or a more complex counter bore and drilled hole, or even a threaded hole.
- Rib: The *Rib* command is used to add material by sweeping a profile along a center curve.
- Slot: *Slots* are used to remove material by sweeping a profile along a center curve.
- Advanced Sketch-Based Features toolbar: This toolbar contains the commands, *Solid Combine* and *Stiffener*. *Stiffener* is used to create relatively thin supporting structures for your part.
- Multi-sections Solid: *Multi-sections Solid* is used to create or add material by connecting several sketches.

- Remove Multi-sections Solid: *Remove Multi-sections Solid* is used to remove material by connecting several sketches.

The Pads toolbar

The *Pads* toolbar, along with the simple *Pad* command, contains advanced commands for creating pads. The *Pads* toolbar reading from left to right contains the following commands.

- Pad: This command is used to create a simple pad.
- Drafted Filleted Pad: This command is used to create a pad that has angled sides and rounded edges.
- Multi-Pad: This command is used to create pads of varying depth all within a single sketch.

Tutorial 3.1 Start: Part Modeled

The part modeled in this tutorial is shown on the right. The entire part is created exclusively using the commands located in the *Pads* toolbar and their supportive sketches.

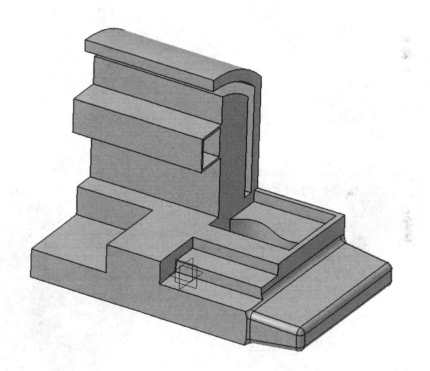

Section 1: Creating Multi-Pads

1) Open a **New…** part drawing and name is *Pad*.

2) Save your drawing as *T3-1.CATPart*.

3) Enter the **Sketcher** on the **xy plane**. Deactivate the Grid's Snap to Point and Display.

4) Before you start to draw make sure your length units are set to millimeters. (**Tools** – **Options...** – **Parameters and Measures**)

5) Draw and **Constrain** the following **Rectangle**.

200

150

75

100

6) **Exit** the *Sketcher*.

7) Pull out the *Pads* toolbar .

8) **Pad** the sketch to a length of **20 mm**. Select **OK**. In the specification tree, double click on Pad.1. The Pad Definition window will appear. If needed you could change the pad features. Select **Cancel**.

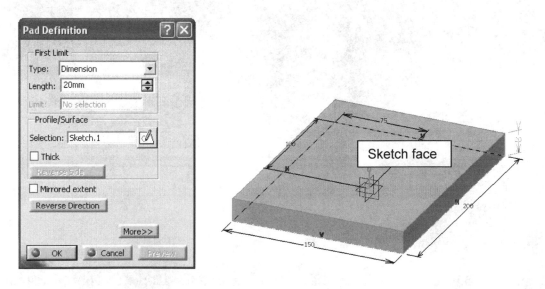

Sketch face

9) Enter the **Sketcher** on the top face of the block.

10) **3D Project Elements** the face of the block on to the sketch plane as a **Construction Element**.

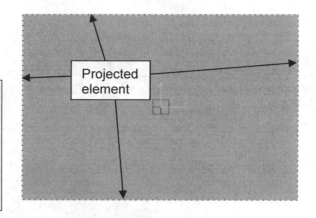

Why? Why are we using construction elements? The construction elements projected in step 10) will help us locate the Rectangles created in step 11). We will snap the Rectangles to the construction element. You know that you have snapped the Rectangle to the construction element if you create a coincident constrain ⊙.

11) Draw and **Constrain** the following **Rectangles**. Make sure to snap exactly to the corner. Add enough constrain to create an iso-constrained (green) sketch.

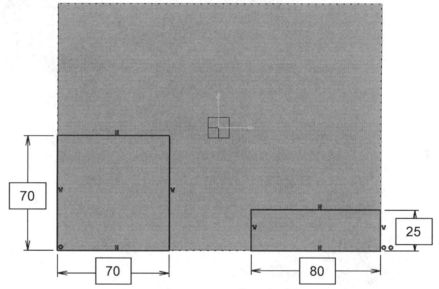

12) Add the following **Profiles** to the sketch. Make sure not to draw overlapping lines.

(Problem? Is your sketch <u>over constrained</u> (pink)? Look at the coincident constraints. These are usually the culprit. CATIA may create a coincident constraint between two distance corners. Delete this constraint and see if the sketch becomes iso-constrained (green).)

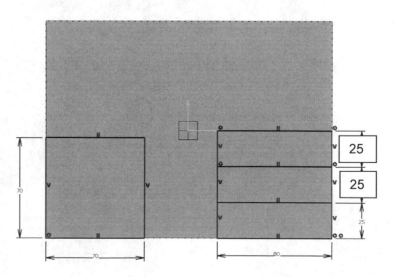

13) **Exit** the Sketcher and select the **Multi-Pad** 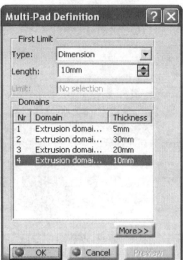 icon. In the *Multi-Pad Definition* window click on one of the Extrusion domains in the Domains field. Notice that a feature will highlight in blue on the part. Type in that feature's length and click on the next, etc... The pad lengths are as follows. The lone rectangle is **5 mm**, the corner rectangle in the group of three is **10 mm**, the next is **20 mm**, and the last is **30 mm**.

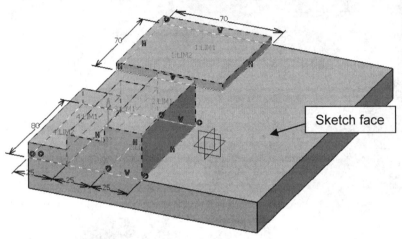

Sketch face

14) **Save** your part.

Section 2: Advanced Pads

2-1: Adding and removing elements of a sketch.

CATIA allows you to *Pad* only a portion of a sketch. The user may manually add or remove elements from the profile definition.

1) Enter the **Sketcher** on the top face of the main pad as shown in the figure above.

2) **3D Project Elements** the top face of the main pad on to the sketch plane as a **Construction Element**.

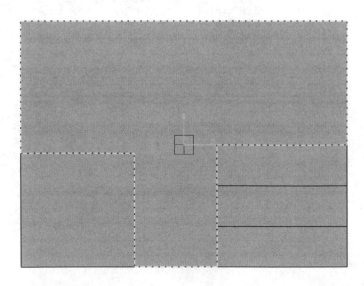

Why? Just like before, we are projecting the construction elements to help locate the profiles created in the next step. We will snap the profile to the construction element and obtain the coincident constraints.

3) Draw the two **Profiles** shown in the figure. The dimensions of the small rectangle is not important.

4) **Exit** the Sketcher and deselect all.

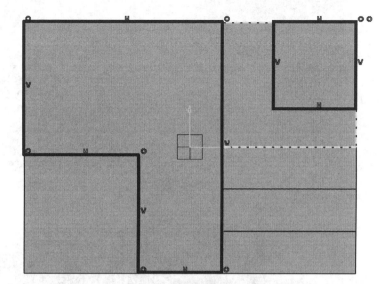

5) Select the **Pad** icon. In the *Pad Definition* window, right click on the `Selection:` field and click on **Go to profile definition**. In the *Profile Definition* window, activate the **Sub-element** toggle and click the **Add** button and select the profile indicated in the figure. Select **OK**. In the *Pad Definition* window, set the length to **30 mm** and apply the pad.

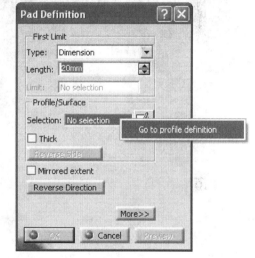

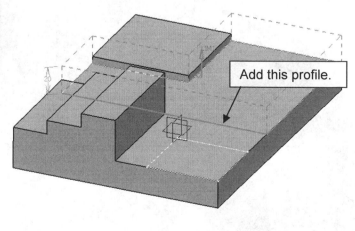

Add this profile.

2-2: Padding using an open profile.

CATIA allows you to *Pad* an open profile as long as it is bounded by the part.

1) Right click on **Sketch.3** (the last sketch) in the specification tree and select **Hide/Show**.

2) Enter the **Sketcher** on the top face of the main pad.

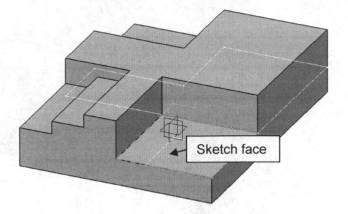

3) **3D Project Elements** the top face of the main pad on to the sketch plane as a **Construction Element**.

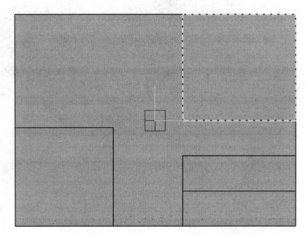

4) Draw and **Constrain** the closed **Profile** shown in the figure.

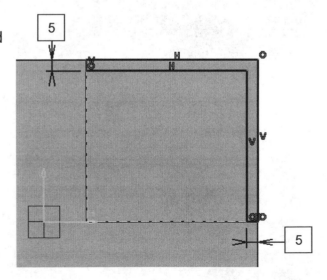

5) **Exit** the *Sketcher* and **Pad** 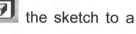 the sketch to a length of *30 mm*.

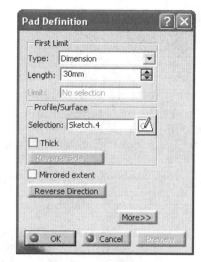

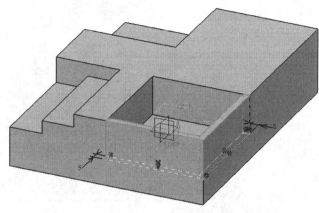

6) Enter the **Sketcher** on the top face of the main pad.

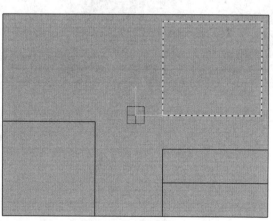

Sketch face

7) **3D Project Elements** the top face of the main pad on to the sketch plane as a **Construction Element**.

8) Draw a **Spline** similar to that shown in the figure.

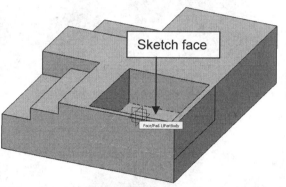

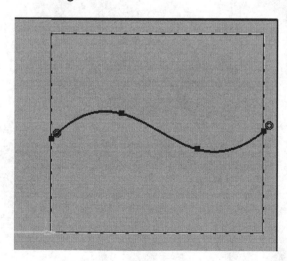

9) **Exit** the *Sketcher* and select the **Pad**  icon. Set the length to **20 mm** and select the **Preview** button. Select the **Reverse Side** button and **Preview** again (you may have to click on the screen to reactivate the preview button). Notice the difference. Apply the *Pad* when it looks like the figure.

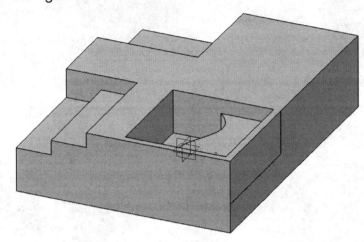

10) Save your part.

2-3: Padding to a surface.

CATIA allows you to *Pad* up to a surface or offset from the surface. If an offset option is used, the end of the *Pad* takes on the same shape as the surface.

1) Rotate the part and enter the **Sketcher** on the face indicated in the figure.

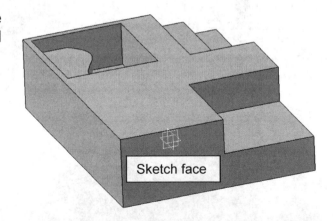

Sketch face

2) **3D Project Elements** the sketch face on to the sketch plane as a **Construction Element**.

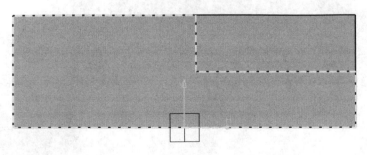

3) Draw and **Constrain** the following **Profile**. Note that the centers of the two arcs are coincident (the same) and the two centers are coincident with the *v* axis. This constraint may be applied after the two three-point-arcs are drawn.

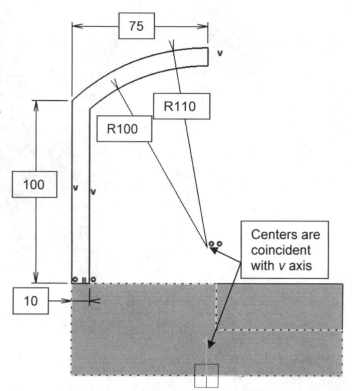

4) **Exit** the *Sketcher* and **Pad** the sketch to a length of **120 mm**. You may have to reverse the direction.

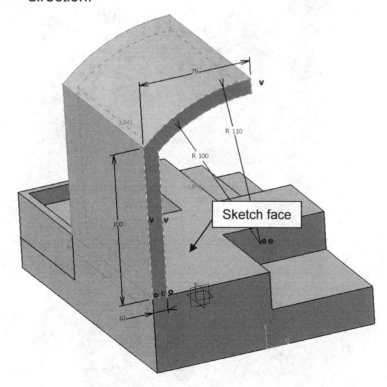

5) Enter the **Sketcher** on the top flat face of the part.

6) **Cut Part By Sketch Plane** .

7) **3D Project Elements** the sketch face on to the sketch plane as a **Construction Element**.

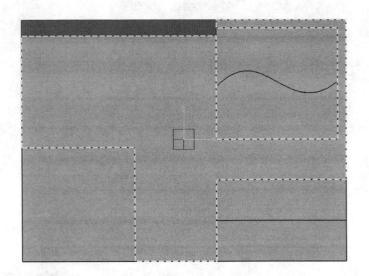

8) Draw and **Constrain** the **Rectangle** shown in the figure.

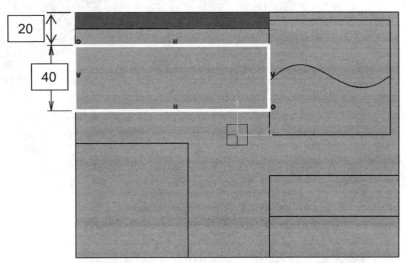

9) **Exit** the Sketcher and **Pad** the sketch using the option of **Up to surface**. Activate the `Limit:` field and select the surface indicated in the figure and set the `Offset:` to *-10 mm*.

10) **Save** your part.

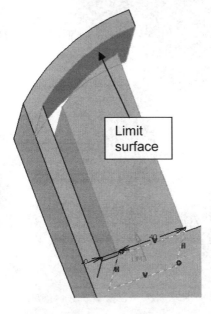

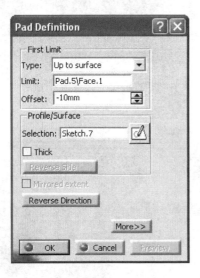

2-4: Padding with thickness.

CATIA gives you the option of padding a sketch with thickness. In simple terms, it means that you can easily create a thin walled section.

1) Rotate the part and enter the **Sketcher** on the face indicated in the figure.

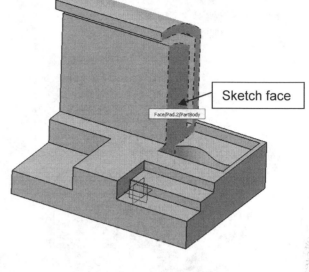

Sketch face

Face/Pad.2/PartBody

2) **3D Project Elements** the left edge of the sketch face on to the sketch plane as a **Construction Element**.

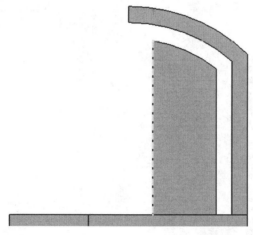

3) Draw and **Constrain** the **Rectangle** shown in the figure.

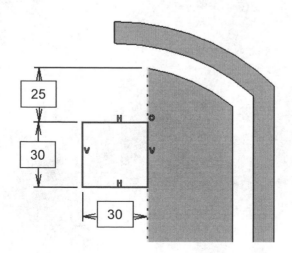

25

30

30

4) **Exit** the Sketcher and select the **Pad** 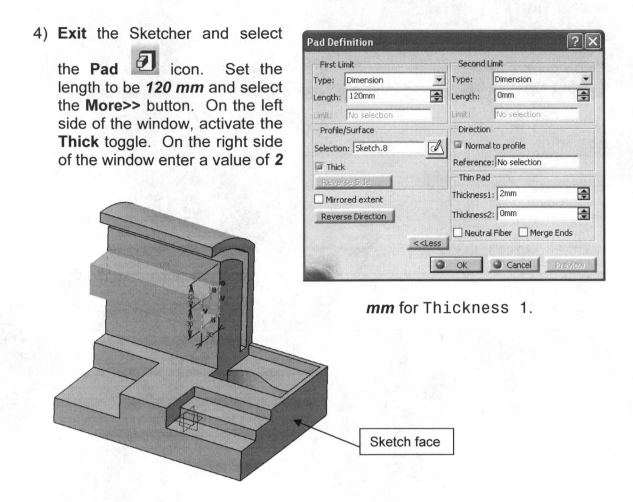 icon. Set the length to be *120 mm* and select the **More>>** button. On the left side of the window, activate the **Thick** toggle. On the right side of the window enter a value of *2*

mm for Thickness 1.

Sketch face

Section 3: Creating Drafted Filleted Pads

1) Enter the **Sketcher** on the right face of the part as shown in the figure above.

2) Draw and **Constrain** the **Rectangle** shown in the figure.

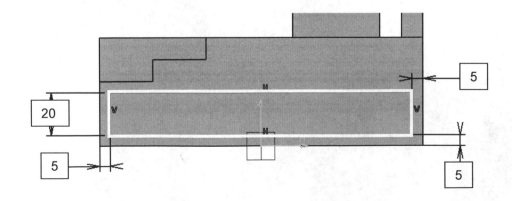

3) Exit the Sketcher and select the **Drafted**

Filleted Pad 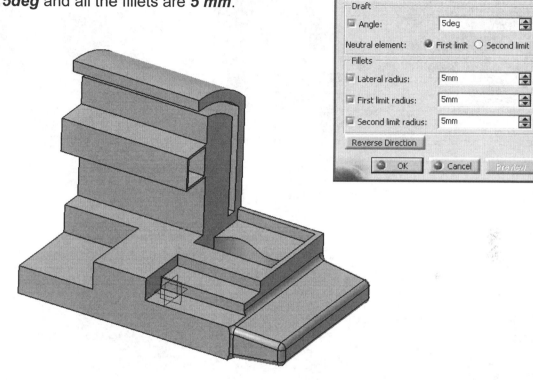 icon. Set the Length: to be *50 mm*. Activate the Limit: Field and select the sketch face. The draft angle is *5deg* and all the fillets are *5 mm*.

Drafted Filleted Pad Definition [?] [X]

First Limit

Length: 50mm

Second Limit

Limit: Pad.7\Face.2

Draft

☑ Angle: 5deg

Neutral element: ● First limit ○ Second limit

Fillets

☑ Lateral radius: 5mm

☑ First limit radius: 5mm

☑ Second limit radius: 5mm

Reverse Direction

● OK ● Cancel Preview

4) **Save** your part.

NOTES:

Chapter 3:
PART DESIGN

Tutorial 3.2: Sketch-Based Features: Pockets, Holes & Slots

Featured Topics & Commands

Prerequisite Knowledge & Commands

- The *Sketcher* workbench and associated commands
- Editing a preexisting sketch
- Pads
- Rotating the part using the mouse or compass

The Pockets toolbar

The *Pockets* toolbar, along with the simple *Pocket* command, contains advanced commands for creating pockets. The *Pockets* toolbar, reading from left to right, contains the following commands.

- Pocket: This command is used to create a simple pocket.
- Drafted Filleted Pocket: This command is used to create a pocket that has angled sides and rounded edges.
- Multi-Pocket: This command is used to create pockets of varying depth all within a single sketch.

Holes

The *Hole*  command (located in the *Sketch Based Features* toolbar) is used to create holes of different configurations. For example, this command could be used to create a simple blind hole or a more complex counter bore and drilled hole, or even a threaded hole. The *Hole* command is initiated in the *Part Design* workbench without a preexisting sketch. However, the command automatically creates a sketch; therefore, it is still a sketch based feature.

Slots

The *Slot* command (located in the *Sketch Based Features* toolbar) is used to remove material by sweeping a profile along a center curve. The profile may be opened or closed depending on the options chosen. The center curve should not be made of multiple geometric elements.

Tutorial 3.2 Start: Part Modeled

The part modeled in this tutorial is shown on the right. It is created from one main pad and then modified using the *Pocket, Hole* and *Slot* commands. Several different *Pocket* and *Hole* options are explored.

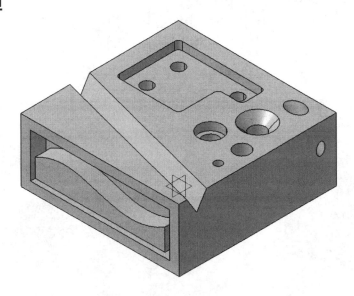

Section 1: Creating Pockets

1) Open a **New... Part** and name it **Pockets**.

2) **Save** your part as **T3-2.CATPart.**

3) Enter the **Sketcher** on the **xy plane**.

4) Draw and **Constrain** the following **Rectangle**.

5) **Exit** the *Sketcher* and **Pad** 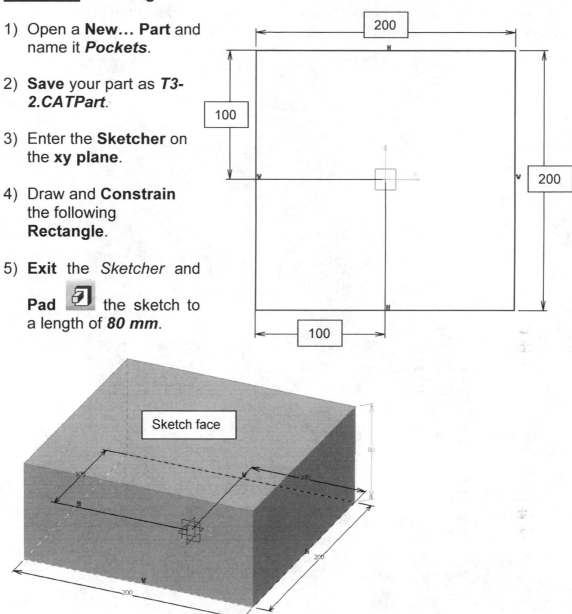 the sketch to a length of **80 mm**.

6) Enter the **Sketcher** on the top face of the block.

7) Draw and **Constrain** the following sketch. All the **Corner** (fillet) radii are **5 mm** and all the hole diameters are **15 mm**. To avoid unwanted constrains, hold the **SHIFT** key down when drawing the circles.

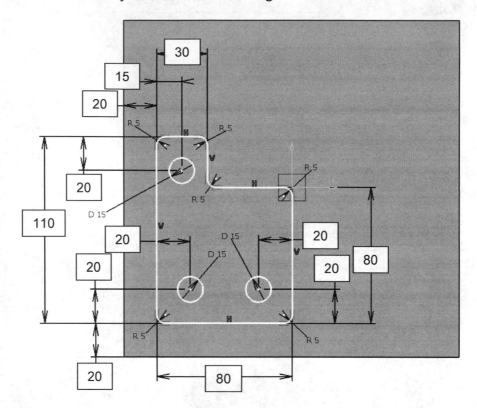

8) Pull out the **Pockets** toolbar

9) **Save** your part.

10) **Exit** the *Sketcher* and select the **Multi-Pocket** icon. Set the depth of the inside holes to *80 mm* and the depth of the outside profile to be *10 mm*. Remember when you click on one of the Extrusion domains it will highlight the respective feature on the part.

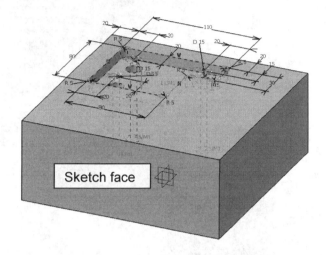

Sketch face

11) Enter the **Sketcher** on the left face of the part as shown in the figure.

12) **3D Project Elements** the sketch face on to the sketch plane as a **Construction Element**.

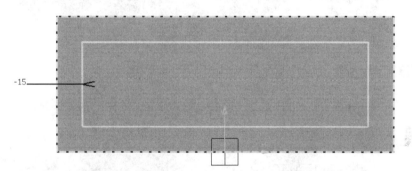

13) Deactivate the **Standard/Construction Element** icon. Select the projection and then select the **Offset** icon. Click your mouse to the inside of the projection. Set the offset dimension to be *-15 mm*. The offset rectangle should be a **Standard Element**.

14) **Exit** the *Sketcher* and select the **Pocket** 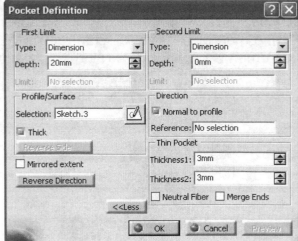 icon. In the *Pocket Definition* window, select the **More>>** button. Activate the **Thick** toggle. Set the pocket depth to be *20 mm* and both thickness' to be *3 mm*. Set and preview one thickness at a time to see which direction is effected by both.

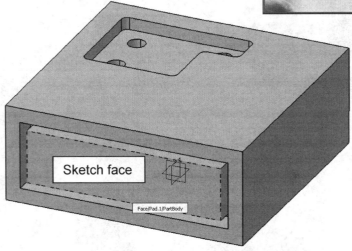

15) Enter the **Sketcher** on the left face of the part.

16) **3D Project Elements** the inside portion of the sketch face on to the sketch plane as a **Construction Element**.

17) Draw a **Spline** similar to the one shown in the figure.

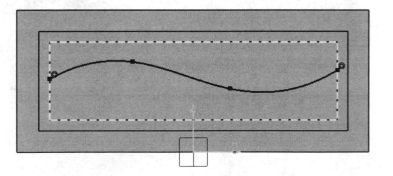

18) **Exit** the *Sketcher* and

Pocket 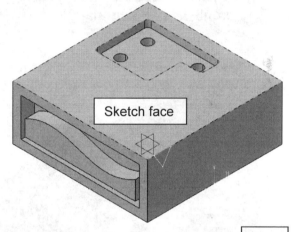 the sketch to a depth of **20 mm**. Use the *Reverse Side* button if necessary to achieve the same results as shown in the figure.

19) **Save** your part.

Section 2: Creating Slots.

1) Enter the **Sketcher** on the top face of the part.

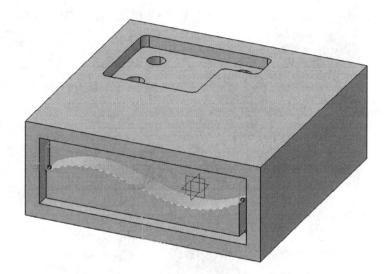

Sketch face

2) **3D Project Elements** the sketch face on to the sketch plane as a **Construction Element**.

3) Draw and **Constrain** a **Standard Element Line** as shown in the figure.

4) **Exit** the Sketcher and deselect all.

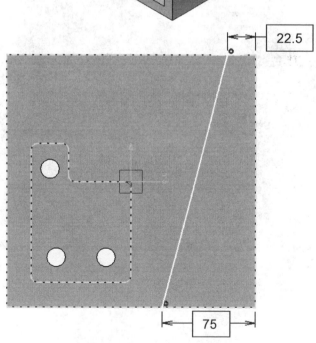

22.5

75

5) Enter the **Sketcher** on the right face of the part.

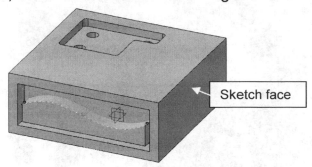

Sketch face

6) **3D Project Elements** the top edge of the sketch face on to the sketch plane as a **Construction Element**.

7) Draw and **Constrain** the following **Profile**. The profile is a closed triangle.

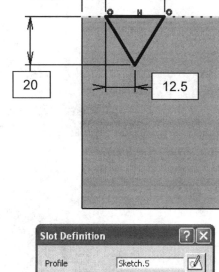

10   25

20   12.5

8) Select the **Slot**  icon. Use the triangle as the Profile: and the line as the Center curve:.

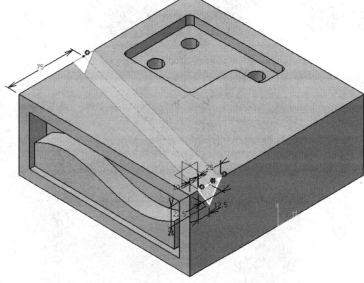

9) **Save** your part.

Section 3: Creating Holes.

1) Select the **Hole** 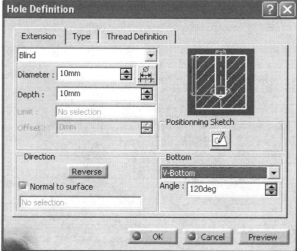 icon and then select the top face of the part. Select the **Extension** tab in the *Hole Definition* window. Create a **Blind** hole with a Diameter: of *10 mm*, a Depth: of *10 mm*, and a **V-Bottom**. Click on the **Positioning Sketch** icon and **Constrain** the center of the hole as shown in the figure.

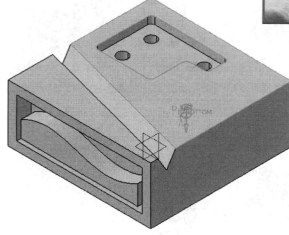

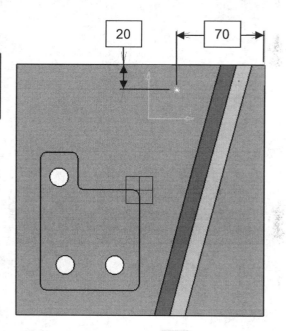

2) **Exit** the Sketcher and select **OK**.

3) Create another *20 mm* Diameter: through (**up to last**) **Hole** 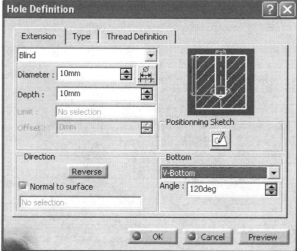 on the top face of the part. The hole should be constrained as shown in the figure.

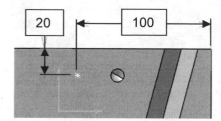

4) Create a counterbored **Hole** 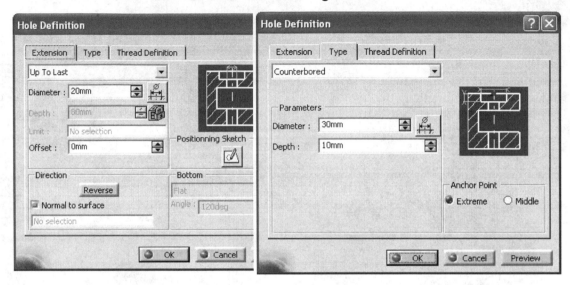 with the following parameters. Select the **Type** tab to change the hole type from simple to **Counterbored**.
- Counterbore Diameter: **30 mm**
- Counterbore Depth: **10 mm**.
- Drill Diameter: **20 mm** and it is a through hole.

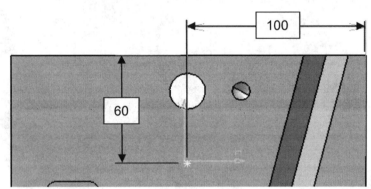

5) Create a countersunk **Hole** 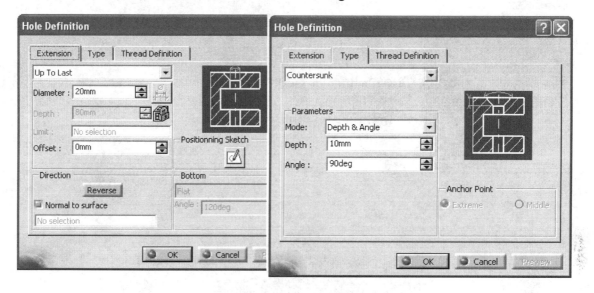 with the following parameters.
   - Mode: **Depth & Angle**
   - Countersunk Depth: **10 mm**
   - Angle: **90deg**.
   - Drill Diameter: **20 mm** and it is a through hole.

6) **Save** your part.

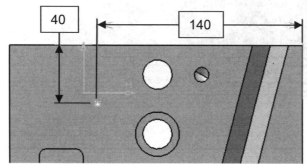

7) Rotate your part so that the back side is visible.

8) Create a reference line that will control the direction of a hole. Select the **Line** 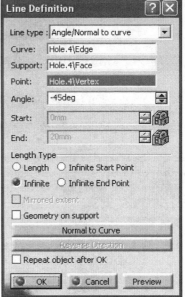 icon. In the *Line Definition* window, fill in the following fields;
   - Line type: **Angle/Normal to curve**
   - Curve: Select the bottom edge of the back face
   - Point: Select the bottom left corner of the surface
   - Support: Select the surface (see fig.)
   - Angle: *-45deg*
   - Length Type: **Infinite**

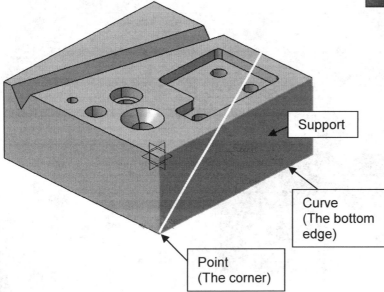

Support

Curve
(The bottom edge)

Point
(The corner)

9) Create a *10 deg* **Tapered** through **Hole** 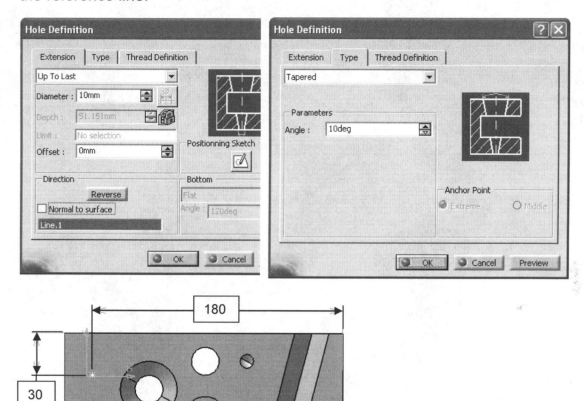 on the top face with a *10 mm* diameter that uses the reference line as its direction. To select the direction, deactivate the **Normal to surface** toggle, activate the field below it and select the reference line.

10) **Save** your part.

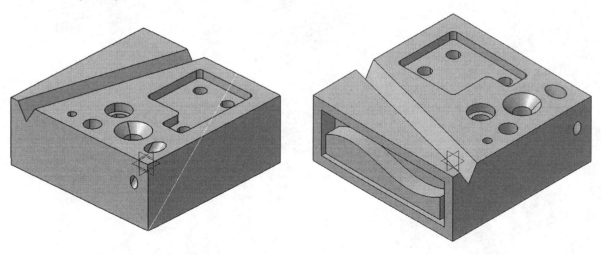

NOTES:

Chapter 3:
PART DESIGN

Tutorial 3.3: Sketch-Based
Features: Shafts & Grooves

Featured Topics & Commands

Prerequisite Knowledge & Commands

- The *Sketcher* workbench and associated commands
- Editing a preexisting sketch
- Pads
- Rotating the part using the mouse or compass

Shafts

A *Shaft* creates a solid by revolving a profile around an axis. Open or closed profiles may be used provided that the profile starts and ends at the axis. Only closed profiles can be used if the profile is away from the axis. The shaft cannot be created if the profile crosses the axis. See the examples shown below.

- Closed or open section may be used when the profile starts and ends on the axis.

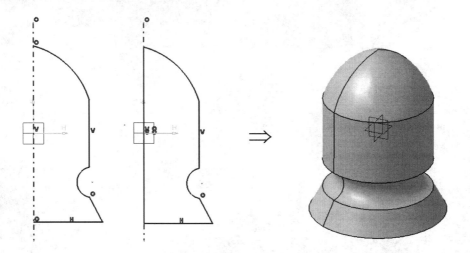

- A closed profile must be used if the sketch does not contact the axis.

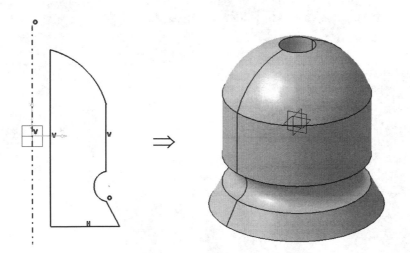

- Profiles that will not produce a shaft.

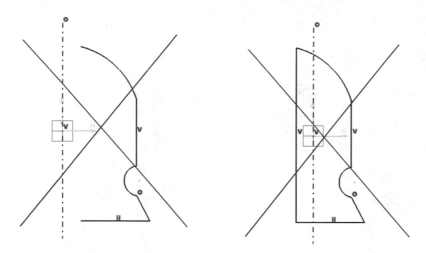

Grooves

A *Groove* 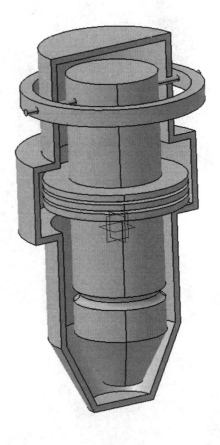 removes material by revolving a profile around an axis. The profile may be open or closed depending on the options chosen in the *Groove Definition* window. But, the profile should never cross the axis. This command is very similar to the *Shaft* command except that it removes material.

Tutorial 3.3 Start: Part Modeled

The part modeled in this tutorial is shown on the right. The main solid is created using a *Shaft*. Details are added using both the *Shaft* and *Groove* commands.

Section 1: Creating Shafts.

1) Open a **New...** part drawing and name it **Shaft.**

2) Save your part as **T3-3.CATPart**.

3) Enter the **Sketcher** on the **yz plane**.

4) Set your length units to be inches (**Tools** – **Options...** – **Parameters and Measures**)

5) Draw and **Constrain** the following **Profile** and **Axis**.

6) **Exit** the Sketcher and **Shaft** the sketch through an angle of **360deg** using the **Sketch Axis** as the Axis Selection.

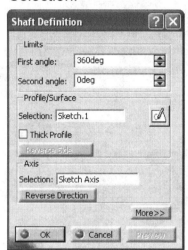

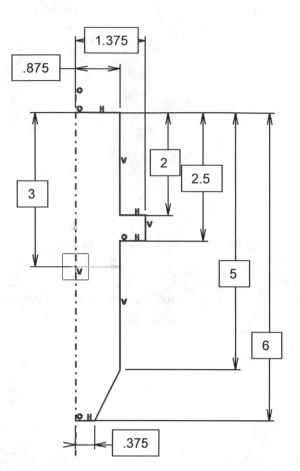

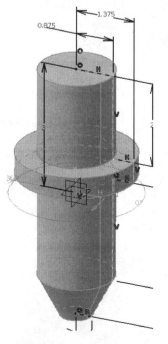

7) Enter the **Sketcher** on the **zx plane**.

8) Use the **3D Project Element** and **Project 3D Silhouette Edges** to project all the edges of the part as **Construction Elements**.

9) Draw an **Axis** that is on and parallel to the V axis.

10) Using the projections as a guide, draw a **Profile** around the outline of the part on the right side of the axis as a **Standard Element**.

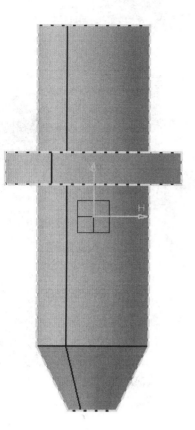

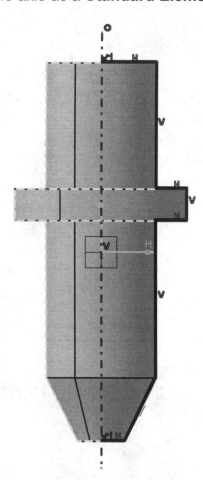

11) Offset the profile to the outside of the part. Select all elements of the profile you created in the previous step. Select the **Offset** icon. Activate the Offset: field in the *Value field* using the **TAB** key. Enter **0.5** and hit the **TAB** key again.

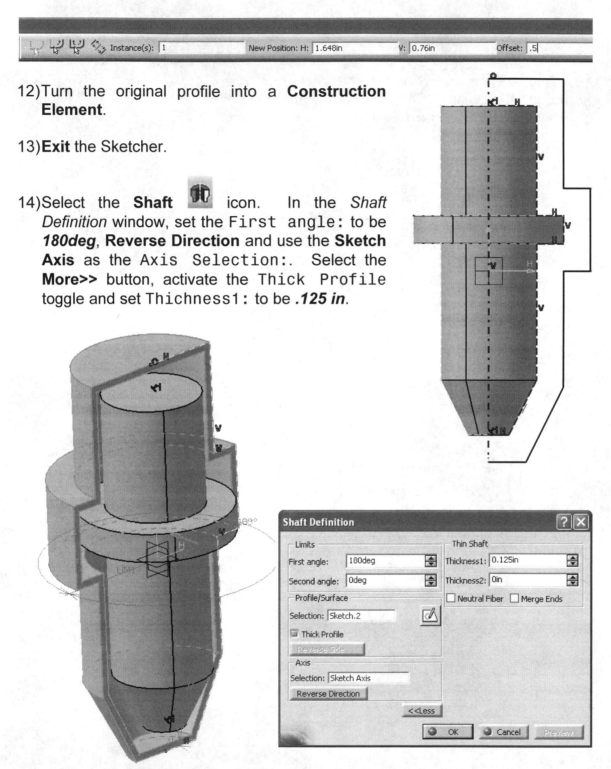

12) Turn the original profile into a **Construction Element**.

13) **Exit** the Sketcher.

14) Select the **Shaft** icon. In the *Shaft Definition* window, set the First angle: to be **180deg**, **Reverse Direction** and use the **Sketch Axis** as the Axis Selection:. Select the **More>>** button, activate the Thick Profile toggle and set Thichness1: to be **.125 in**.

3.3 - 6

15) Enter the **Sketcher** on the **zx plane** and draw and **Constrain** the **Rectangle** shown.

16) **Exit** the *Sketcher* and **Shaft** 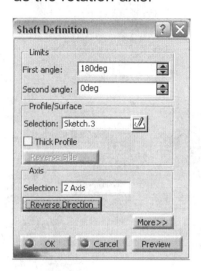 the sketch *180deg* using the Z-axis as the rotation axis.

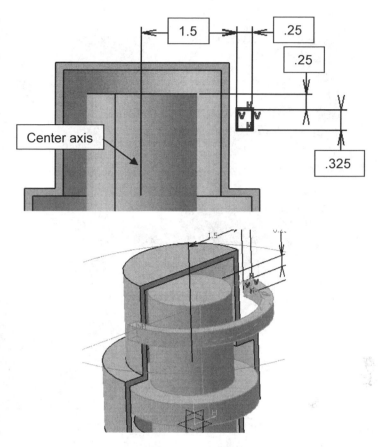

17) Save your part.

Section 2: Creating Grooves.

1) Enter the **Sketcher** on the **zx plane**.

2) Draw and **Constrain** the following **Profile**.

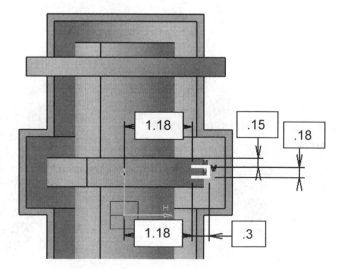

3) **Exit** the Sketcher.

4) Select the **Groove** 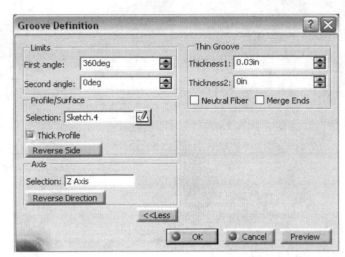 icon. In the *Groove Definition* window, set the First angle: to be ***360deg*** and use the Z-axis as the rotation axis. Select the **More>>** button, activate the Thick Profile toggle and set Thickness1: to be ***0.03 in***. Select **OK**.

5) Double click on **Groove.1** in the specification tree.

6) In the Groove Definition window, set Thickness2: to be ***0.03 in***.

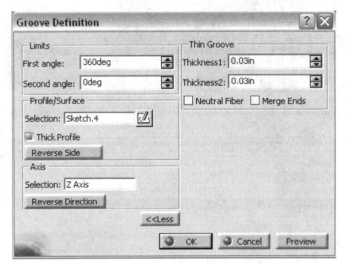

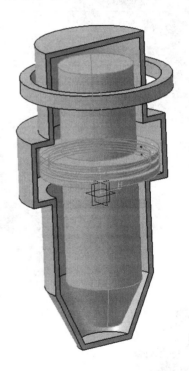

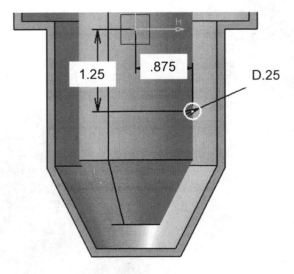

7) Enter the **Sketcher** on the **zx plane**. Draw and **Constrain** the **Circle** shown.

8) **Exit** the *Sketcher* and create a ***360 deg* Groove** 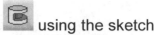 using the sketch.

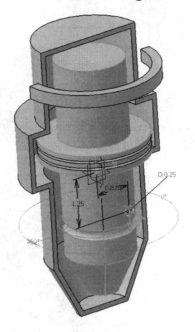

9) Edit **Shaft.3**. In the Shaft Definition window set the rotation angle to be ***360deg***.

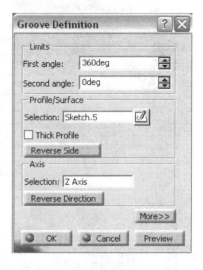

10) Enter the **Sketcher** on the **yz plane**.

11) Draw and **Constrain** the **Circle** shown. The center of the circle lies on the center axis.

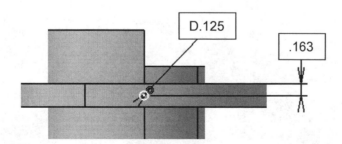

12)**Exit** the *Sketcher* and **Pad** 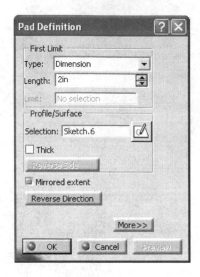 the circle *2* inches in both directions (Mirrored extent).

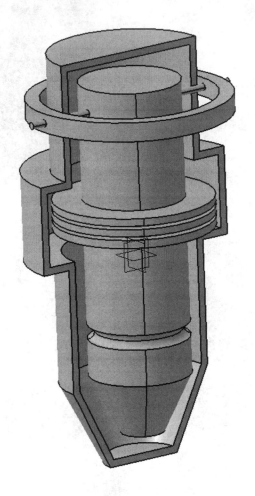

13)Save your part.

Chapter 3:
PART DESIGN

Tutorial 3.4: Sketch-Based Features: Advanced commands

Featured Topics & Commands

Prerequisite Knowledge & Commands

- The *Sketcher* workbench and associated commands
- Editing a preexisting sketch
- Pads
- Rotating the part using the mouse or compass

Multi-sections Solid

A *Multi-sections Solid* is generated by two or more planar sections that are swept along a spine. A *Multi-sections Solid* can add material or remove material. The planar sections may also be connected with guidelines which act as boundaries in which the *Multi-sections Solid* is contained. Direction arrows show the orientation of the *Multi-sections Solid* and can be used to change the orientation. *Multi-sections Solids* are very useful in creating transitions between two existing solids.

<u>Note:</u> It seems that in R17 the spine curve must be created before the Multi-sections Solid. Otherwise CATIA will not allow you re-enter the *Multi-sections Solid* definition window and select the curve.

Ribs and Slots

A *Rib* is a profile swept along a center curve (open or closed) to create a solid feature. *Ribs* are useful when you need to sweep a profile from one surface to another. A *Rib* may also be used to create a pipe like feature by sweeping a profile along a center curve.

A *Slot* is a profile swept along a center curve (open or closed) to remove material from a solid.

The center curve used to create *Ribs* and *Slots* do not have to be drawn to the feature's true length. Walls of existing parts may be used as boundaries.

Stiffener

A *Stiffener* is a brace that is added to a wall or other structures that need to be strengthened to prevent possible breakage. It is often used in castings and molded plastics.

Tutorial 3.4 Start: Part Modeled

The part created in this tutorial is a representation of a boat. Notice the sloping surface of the boats hull (Multi-sections Solid), the thin support members (stiffeners), and the ribs and slots that are part of the outer shell.

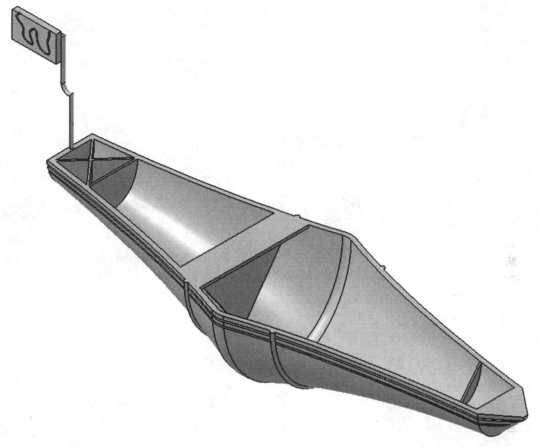

Section 1: Multi-sections Solid

1.1 – Creating a Simple Loft

1) Open a **New… Part** drawing and name it **Boat.**

2) Save your part as **T3-4.CATPart**.

3) Enter the **Sketcher** on the **zx-plane.**

4) Set your length units to millimeters.

5) Draw and **Constrain** the following **Ellipse**. The ellipse's center coincides with the origin and the axes are along the V and H coordinate directions.

6) Draw a **Line** that cuts the ellipse in half as shown.

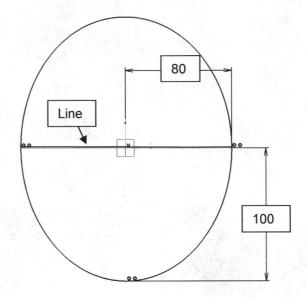

7) **Quick Trim** 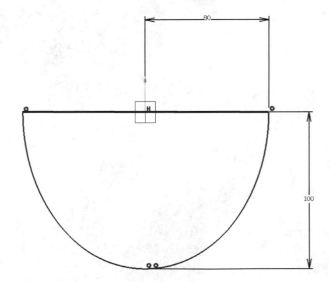 the top portion of the ellipse.

8) **Exit** the Sketcher and deselect all.

9) Create a new **Plane**  by offsetting the **zx-plane** by **50 mm**. In the *Plane Definition* window, activate the **Repeat object after OK** toggle. After you select **OK** an *Object Repetition* window will appear. Enter **4** instances.

10) Create the same number of new planes on the other side of the sketch as shown.

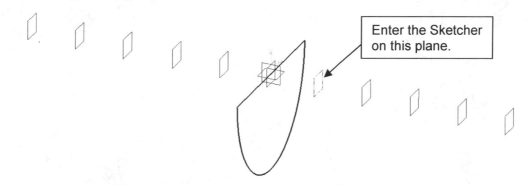

Enter the Sketcher on this plane.

11) Enter the **Sketcher** on the first new plane to the right of the sketch (Plane.1 in your tree). See above figure.

Center point

12) Project Sketch.1 onto the new plane using **3D Project Elements**. To select the entire sketch all at once, select Sketch.1 in the specification tree.

13) Scale the projection by 80% using the following procedure. Select all elements of the projection. Select the **Scale** icon. Select the scaling center point as shown in the figure, activate the **Duplicate mode**, and enter **0.8** in the Value: field.

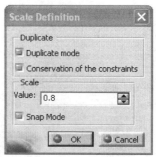

14) Turn the original projection into a **Construction Element**.

15)**Exit** the *Sketcher*.

16)Repeat the above procedure to create 4 more sketches all scaled 80% from that of the previously sketch.

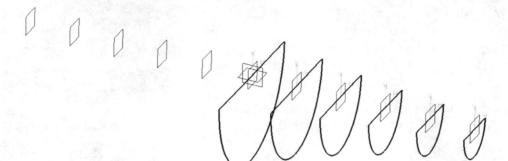

17)Repeat for the other side.

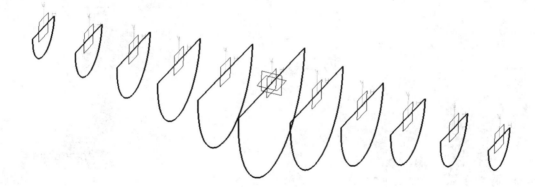

18)Save your part.

19) Create a Multi-sections Solid using the 6 sketches on the right. Deselect all elements. Select the **Multi-sections Solid**

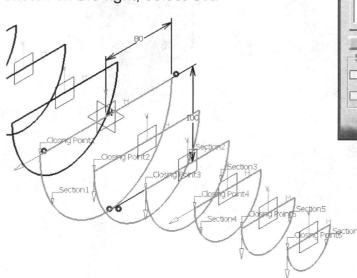 icon. Select the middle (biggest) sketch and then each sketch in order from left to right. After you have selected the last sketch on the right, select **OK**.

20) Repeat for the other side. However, the largest sketch now must be selected in the specification tree (this is because it is already part of a solid). The largest sketch which should be **Sketch.1**. The other sketches may still be selected manually with the mouse.

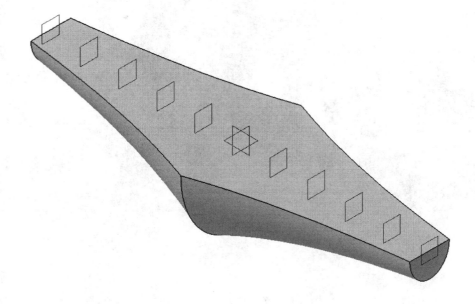

21)Locate the *Dress-Up Features* toolbar

22)**Shell** the solid to a thickness of **5 mm** removing the top face.

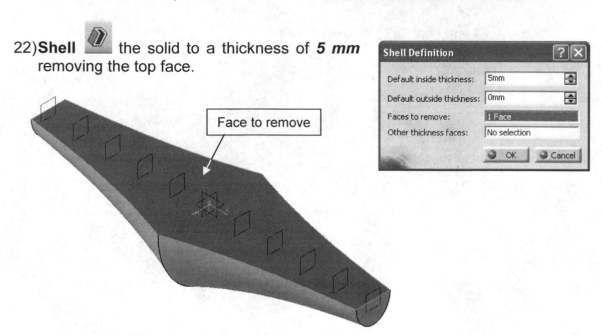

Face to remove

Shell Definition

Default inside thickness:	5mm
Default outside thickness:	0mm
Faces to remove:	1 Face
Other thickness faces:	No selection

OK   Cancel

23)**Edge Fillet** the inside and outside joint between the 2 halves of the boat with a radius of **20 mm**.

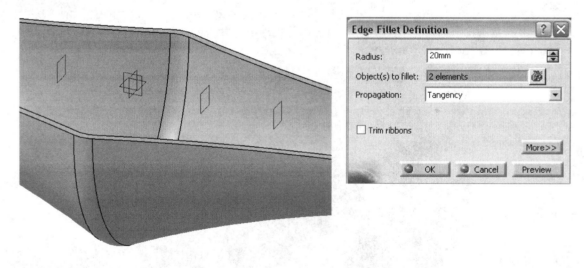

Edge Fillet Definition

Radius:	20mm
Object(s) to fillet:	2 elements
Propagation:	Tangency

☐ Trim ribbons

More>>

OK   Cancel   Preview

24)Save your part.

1.2 – Relimiting a Multi-sections Solid

By default a Multi-sections Solid will begin and end with the planar section chosen. However, if you use guidelines or a spline to direct your Multi-sections Solid, you can relimit it to begin or end with these curves.

1) Offset a **Plane** from the right end of the boat by **30 mm**.

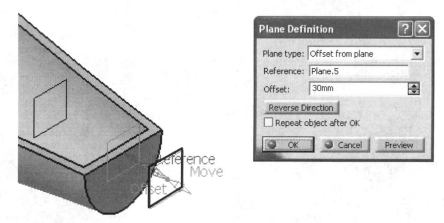

2) Enter the **Sketcher** on the newly offset plane.

3) **3D Project Elements** the end of the boat onto the sketch plane as a **Standard Element**.

4) Scale the projection by a value of **0.1** using the same procedure as before. Turn the original projection into a **Construction Element**.

5) **Exit** the *Sketcher* and re-enter the **Sketcher** on the **yz-plane**.

6) **3D Project Elements** the end of the boat onto the sketch plane as a **Construction Element**.

7) Draw and **Constrain** the **Spline** shown.

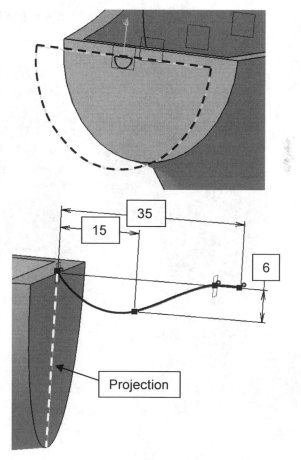

8) **Exit** the *Sketcher* and **Multi-sections Solid**  the end of the boat with the scaled version of the boat end. Select the sketch used to create the end of the boat. This will be located in **Multi-sections Solid.1** the last sketch (**Sketch.6**) and then the small scaled sketch.

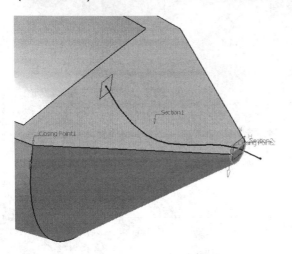

(Having Trouble? If the loft looks funny, like shown below, edit *Multi-sections Solid.3* and change the direction of *Close point 2* by clicking on the arrow.)

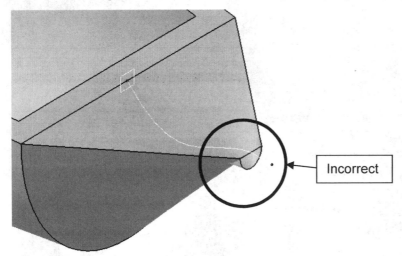

9) Save your part.

10) Double click on **Multi-sections Solid.3**. In the *Multi-sections Solid Definition* window, select the **Spine** tab. Activate the `Spine:` field and select the spline. Select **OK**. Notice the difference.

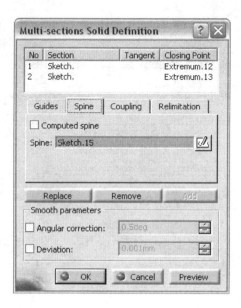

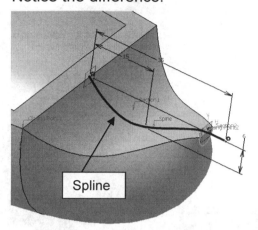

Spline

11) Double click on **Multi-sections Solid.3**. In the *Multi-sections Solid Definition* window, select the **Relimitation** tab and deselect the **Relimited on end section** toggle. Select **OK**. Notice the difference.

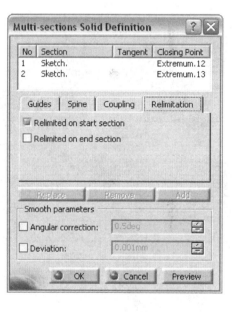

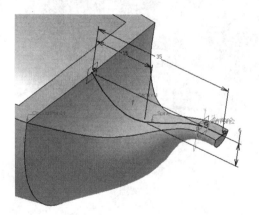

12) Double click on **Multi-sections Solid.3**. Remove the spline and reactivate the **Relimited on end section** toggle.

Section 2: Creating Ribs and Slots

2.1 – Creating a Simple Slot

1) Enter the **Sketcher** on the **xy-plane**.

2) **3D Project Elements** the outer profile of the boat onto the sketch plane. (The projection in the figure is shown black, however the real projection will be yellow.)

3) **Exit** the *Sketcher* and re-enter the **Sketcher** on the **zx-plane**.

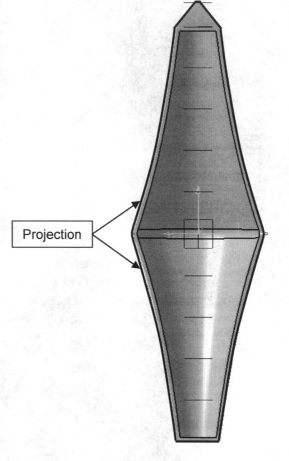

4) Draw and **Constrain** the following **Profile**. Use **Cut Part By Sketch Plane** to get a better view.

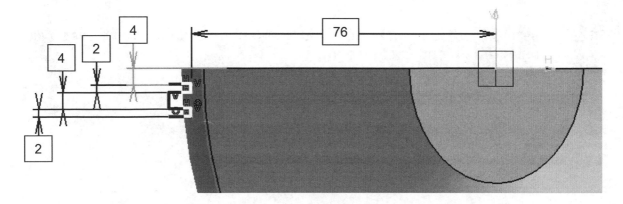

5) **Exit** the Sketcher and create a slot that goes around the boat using the above sketch and the projection. Select the **Slot** 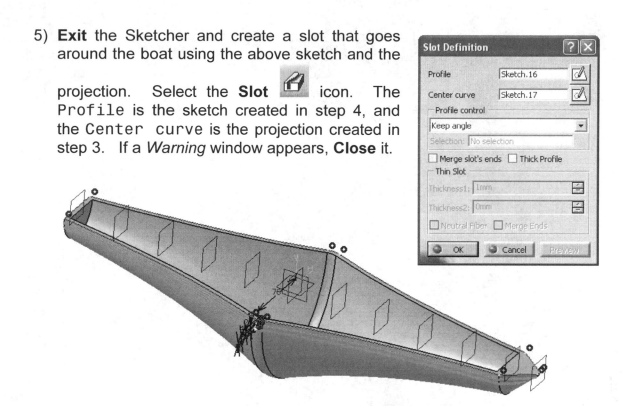 icon. The `Profile` is the sketch created in step 4, and the `Center curve` is the projection created in step 3. If a *Warning* window appears, **Close** it.

6) Enter the **Sketcher** on the **xy-plane**.

7) Draw and **Constrain** the *5 mm* diameter **Circle** as shown.

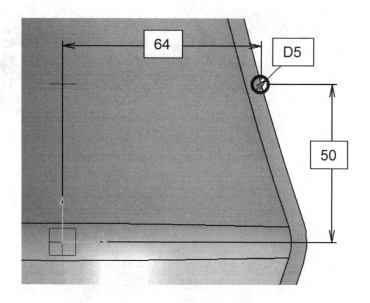

8) **Exit** the *Sketcher* and re-enter the **Sketcher** on the plane indicated in the figure. (The first offset plane created in this tutorial.)

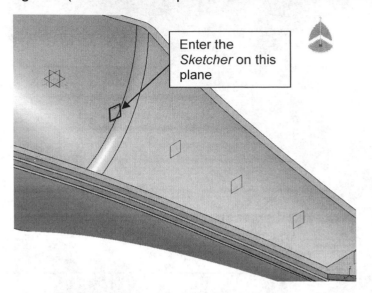

Enter the *Sketcher* on this plane

9) **3D Project Elements** Sketch.2 (the sketch used to create this section of the boat) onto the sketch plane. Use **Cut Part By Sketch Plane** to get a better look.

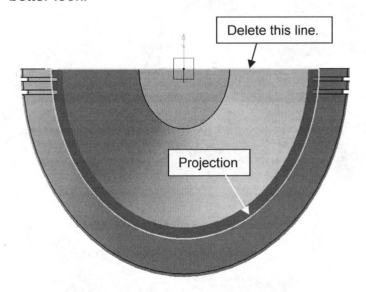

Delete this line.

Projection

10) Isolate the projection and delete the top line of the sketch. (**Insert – Operation – 3D Geometry – Isolate.**)

11) **Exit** the *Sketcher* and create a rib using the above circle and projection. Select the **Rib** icon. Use the circle as the Profile and the projection as the Center curve.

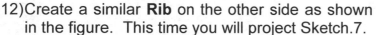

12) Create a similar **Rib** on the other side as shown in the figure. This time you will project Sketch.7.

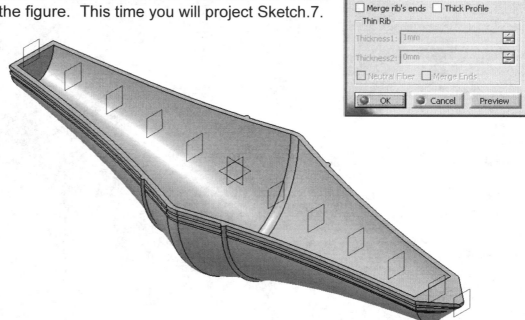

13) Save your part.

14) Enter the **Sketcher** on the top surface of the boat as shown in the figure.

Enter Sketcher on this surface

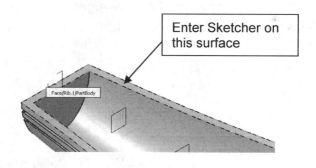

Face/Rib.1/PartBody

15) Draw and **Constrain** the *4 mm* diameter **Circle** shown.

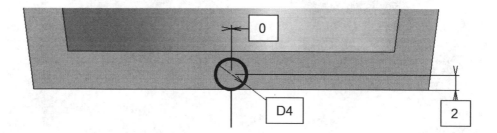

0

D4

2

16) **Exit** the Sketcher and re-enter the **Sketcher** on the **yz-plane**.

17) **3D Project Elements** the top edge of the left end of the boat on to the sketch plane as a **Construction Element**. The projection will appear as a point.

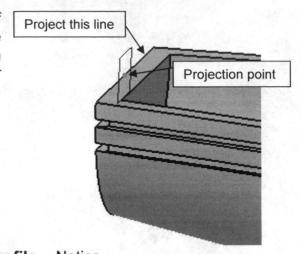

Project this line

Projection point

18) Draw and **Constrain** the following **Profile**. Notice that the bottom of the profile is **Coincident** with the projection point and the top vertical line is **Tangent** with the radius.

19) **Exit** the *Sketcher* and create a **Rib** using the circle as the Profile and the vertical sketch as the Center curve.

Rib Definition

Profile	Sketch.23
Center curve	Sketch.22

Profile control

Keep angle

Selection: No selection

☐ Merge rib's ends ☐ Thick Profile

Thin Rib

Thickness1: 1mm

Thickness2: 0mm

☐ Neutral Fiber ☐ Merge Ends

● OK   ● Cancel   Preview

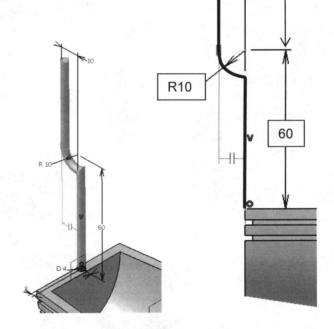

10

50

R10

60

20) Enter the **Sketcher** on the **yz-plane**.

21) Draw and **Constrain** the following **Rectangle**.

22) **Exit** the *Sketcher* and **Pad** 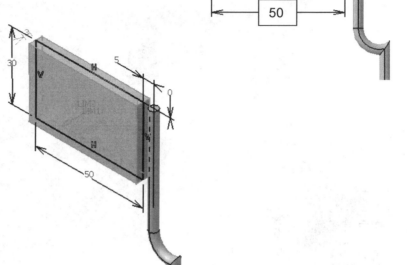 the rectangle to a length of *3 mm* and **Mirror extents**.

23) Enter the **Sketcher** on the top of the rectangle (flag) and draw a circle as shown in the figure. Approximate the size.

24) **Exit** the Sketcher and re-enter the **Sketcher** on the front face of the flag. **3D Project Elements** the front face of the flag onto the sketch plane as a **Construction Element**. Draw a **Spline** that starts and ends at the top of the flag like that shown in the figure. (Do not make it too complex or you may get an error when trying to apply the *Slot*.)

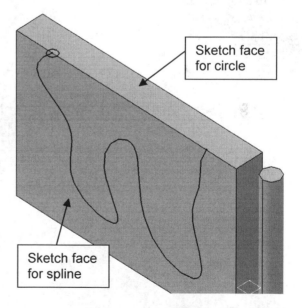

Sketch face for circle

Sketch face for spline

25)**Exit** the Sketcher and create a **Slot** using the circle as the Profile and the spline as the Center curve.

26)Create a **Slot**  using the circle as the Profile and the spline as the Center curve. Activate the **Merge slot's ends** toggle. If a *Warning* window appears, **Close** it.

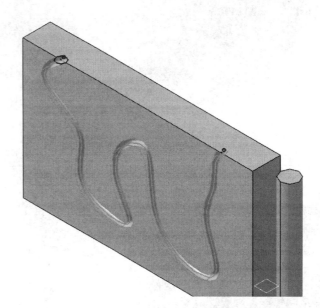

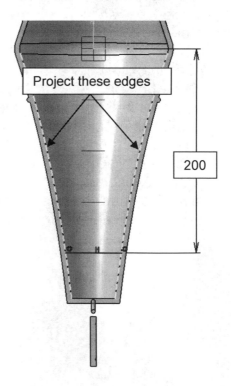

27)Save your part.

Section 3: Creating Stiffeners

1) Enter the **Sketcher** on the **xy-plane**.

2) **3D Project Elements** the left and right inner profiles of the boat on to the sketch plane as **Construction Elements**.

3) Draw and **Constrain** the **Line** shown in the figure.

4) **Exit** the *Sketcher* and create a stiffener that travels from the top down using the following procedure. Deselect all. Select the **Stiffener** 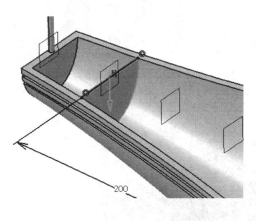 icon. Activate the **From Top** toggle. Set Thickness1 to *2 mm*. Activate the Selection: field and select the line.

5) Enter the **Sketcher** on the **xy-plane**.

6) **3D Project Elements** the inner outline of the boat behind the stiffener as a **Construction Element**.

7) Draw the **Lines** shown.

8) **Exit** the *Sketcher* and create a *2 mm* thick **Stiffener** 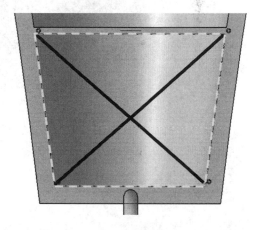 that travels **from** the **top** down using the crossed lines as the profile.

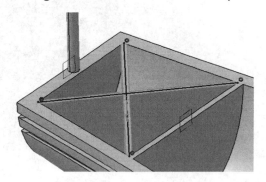

9) On your own, draw and **Pad** 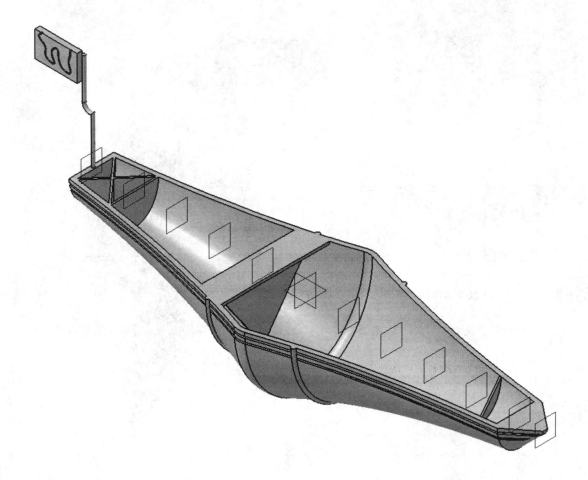 a seat in the boat and support the seat with a **Stiffener** as shown.

10) Save your part.

Chapter 3:
PART DESIGN

Tutorial 3.5: Dress-Up Features: Drafts, Fillets & Chamfers

<u>Featured Topics & Commands</u>

<u>Prerequisite Knowledge & Commands</u>

- The *Sketcher* workbench and associated commands
- Editing a preexisting sketch
- Pads and Pockets
- Rotating the part using the mouse or compass

The Dress-up Features toolbar

Dress-up features allow you to add finishing touches to your part. These commands are not based on a sketch. All parameters of the feature are controlled from the *Definition* window. The commands included in the *Dress-up Features* toolbar are

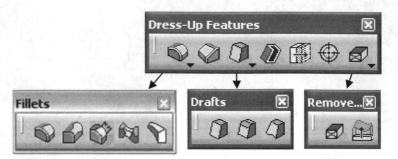

- The Fillets toolbar: The commands contained in the *Fillets* toolbar put radii on sharp corners. Several options are available.
- Chamfer: A *Chamfer* is a bevel created between 2 surfaces.
- The Drafts toolbar: The *Drafts* toolbar contains commands that will place an angle on a surface. Several options are available.
- Shell: The *Shell* command takes a solid part and hollows it out while keeping a given thickness on its sides.
- Thickness: The *Thickness* command adds material uniformly to a surface.
- Thread: Allows you to create threads or taps. The helix thread form is not shown in 3D; however, the information is stored and will show on a 2D drawing.
- The Remove Face toolbar: Allows you to remove or replace a face.

The Drafts toolbar

A *Draft* is an angled face. Material gets added or removed from your part when a draft is applied. The amount of material added and/or removed depends on the draft angle and the location of the neutral surface. Reading from left to right, the commands in the *Drafts* toolbar are

- Draft Angle: Applies a simple draft angle to a surface.
- Draft Reflect Line: This command allows you to draft from a line to a limiting element. It is a complex command and needs to be experimented with to understand its complete usefulness.
- Variable Angle Draft: This command allows you to specify different draft angles to a surface.

When applying a draft, you should keep in mind the following basic draft definitions.

- Pull direction: This is the direction that the draft angle is measured from.

- Draft angle: This is the angle that the draft faces make with the pulling direction.
- Neutral element: This is the element that remains unchanged after the draft command is applied.
- Parting element: This element defines where the draft is to start.
- Limiting elements: Define a region in which the draft should be applied.
- Points: This is used to define the locations where the draft angle varies when using a variable draft.

(Note: A draft angle can never be negative.)

The Fillets toolbar

A *Fillet* is a rounded edge or corner of constant or variable radius that is tangent to the two surfaces that create the edge. Reading from left to right, the commands in the *Fillets* toolbar are

- Edge fillet: This command creates a smooth transition between two surfaces.
- Variable radius fillet: This command allows the fillet to have more than one radius.
- Chordal fillet: This command allows you to control the width of the fillet (distance between the 2 rolling edges) which is also called as chordal length.
- Face-face fillet: This command is used when there is no intersection between the two surfaces or when there are more than two sharp edges between the surfaces.
- Tritangent fillet: This command involves removing one of the 3 surfaces selected.

Tutorial 3.5 Start: Part Modeled

The part created in this tutorial is shown at two different stages of the process. Notice the angled surfaces (drafts), the rounded corners (fillets) and the beveled edges (chamfers).

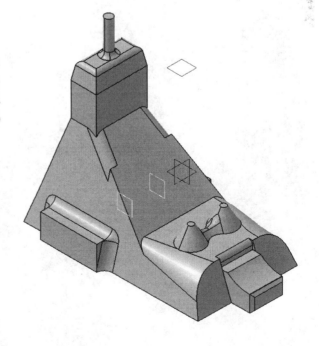

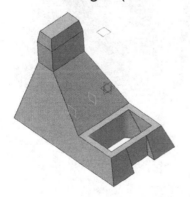

Section 1: Creating Drafts

<u>1.1 - Creating Simple Drafts</u>

1) Enter a **<u>N</u>ew.. Part** drawing and name it **Angled base**.

2) Save the part as **T3-5.CATPart**.

3) Enter the **Sketcher** on the **yz-plane.**

4) Draw and **Constrain** the following **Profile**.

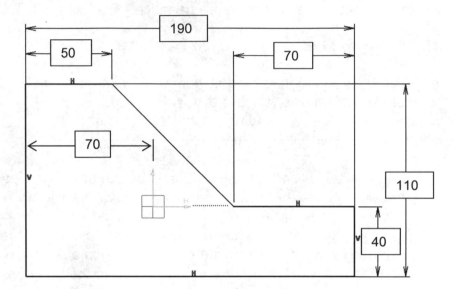

5) **Exit** the *Sketcher* and **Pad** the sketch to a length of **100 mm**.

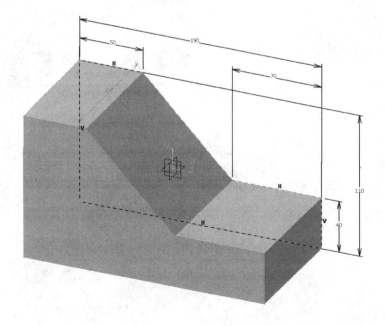

6) Apply a constant **20º Draft Angle** 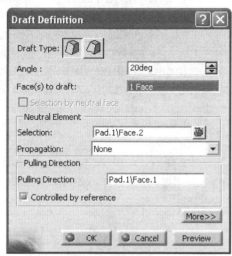 to the right face of the part using the top face as the neutral element. The pulling direction should point up.

Neutral element

Draft face

7) Apply a constant **15º Draft Angle** to the front, back and left side faces of the part using all of the top faces as the neutral elements. The pulling direction should point up.

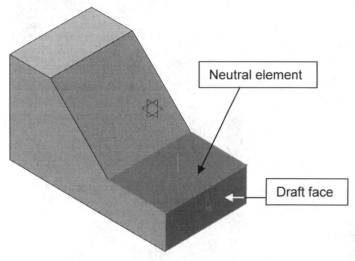

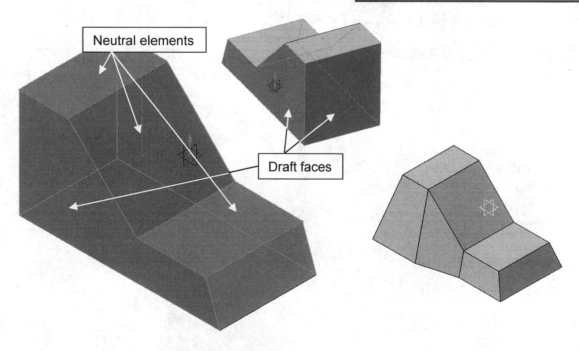

Neutral elements

Draft faces

8) Edit **Draft.2** (the draft applied to the 3 sides) by double clicking on it in the specification tree.

9) Change the neutral faces from the 3 top surfaces to just the lowest one. Activate the *Neutral Element Selection* and click on the very top surface and the angled surface to deselect them.

(Notice: The neutral face indicated in the figure is the only face that remains unchanged.)

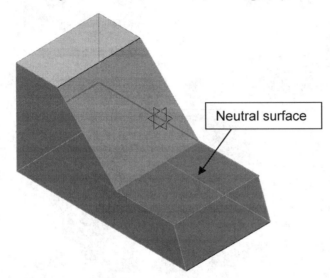

Neutral surface

1.2 - Using a Parting Element:

A *parting element* is a plane, face or surface that cuts the part in two. It is used as a boundary to tell the computer where to stop or change the draft angle.

1) Enter the **Sketcher** on the top face of the part.

2) Use **3D Project Element** to project the top face rectangle onto the sketch plane as a **Standard Element**.

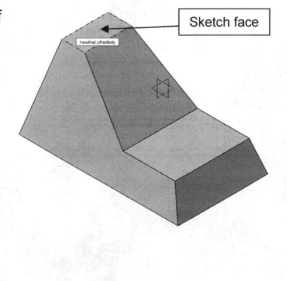

Sketch face

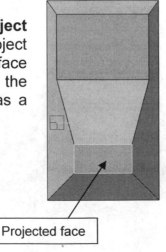

Projected face

3) **Exit** the *Sketcher* and **Pad** 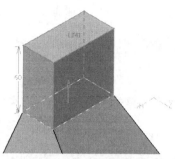 the projection to a length of **60 mm**.

4) Create a **Plane** that is offset **-20 mm** from the top surface of the part.

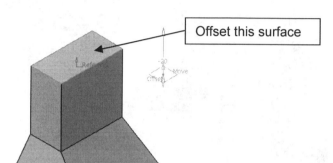

Offset this surface

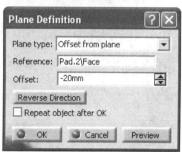

5) Create a **10° Draft Angle** 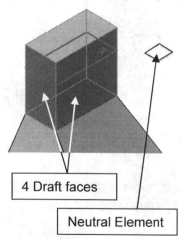 using the offset plane as a parting element. Select the 4 sides of the *Pad* as the Face(s) to draft and the offset plane as the Neutral Element. Select the **More>>** icon and activate the **Parting = Neutral** toggle. The pulling direction should point up. (Note: A curved surface may also be used as a parting element.)

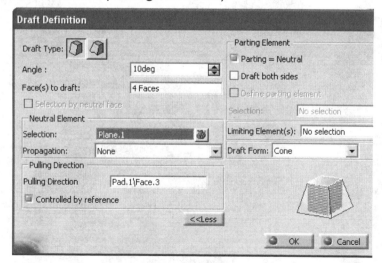

4 Draft faces

Neutral Element

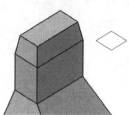

1.3 - Trimming Draft Features

1) Edit **Draft.1** (The first draft created on the end of the part.) by double clicking on it.

2) Select the **More>>** icon in the *Draft Definition* window. Activate the `Limiting Element(s):` right click and select **Create Plane**. Create a plane that is offset *30 mm* from the **yz plane**. Repeat and create a plane that is offset *70 mm* from the **yz plane**. Click on the Orange arrows attached to the offset plane to change their direction to point inward towards each other and apply the draft.

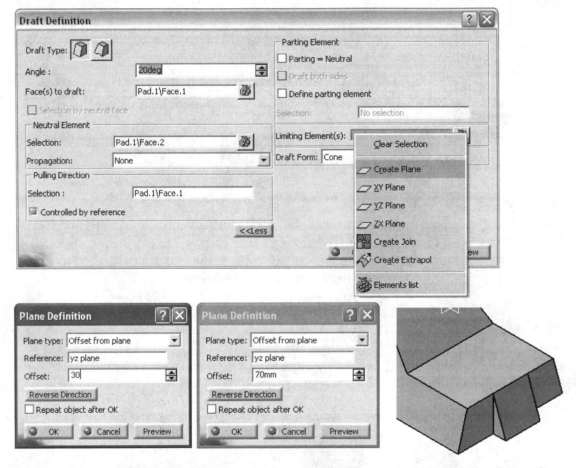

3) Reenter the Draft Definition window and change the direction of the Orange arrows to the outward position and apply the draft and notice the difference.

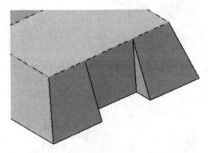

1.4 - Variable Drafts:

A *Variable Angle Draft* allows you to change the draft angle along the surface.

1) Save your part.

2) Enter the **Sketcher** on the top of the base.

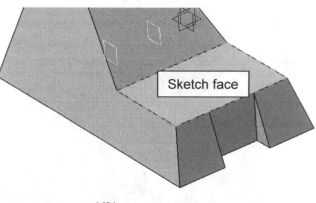

Sketch face

3) Draw and **Constrain** the **Rectangle** as shown. The rectangle is drawn **10 mm** away from all sides.

4) **Exit** the *Sketcher* and **Pocket** the sketch using the **Up to Last** option.

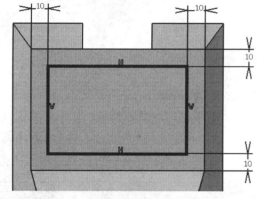

5) Create a **Variable Angle Draft** on the inside left face of the pocket using the top face as the Neutral Element. Once you pick a neutral element, the computer automatically selects 2 points. Add 2 more points by creating 2 planes that are offset from the **yz plane** by **30** and **70 mm**. Double click on the dimensions to change their values.

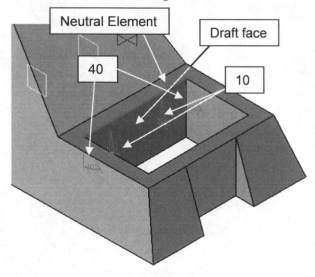

Neutral Element

Draft face

40

10

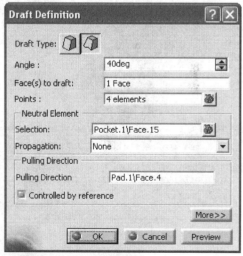

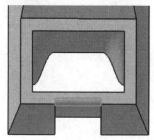

Section 2: Creating Fillets

2.1 - Creating Simple Fillets

1) **Save** your Part.

2) Create a **5 mm** radius **Edge Fillet**  on the top right side edge between the two drafted surfaces.

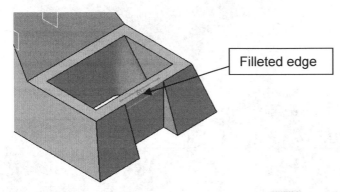

Filleted edge

3) Create a **5 mm** radius **Edge Fillet**  around the very top face of the part.

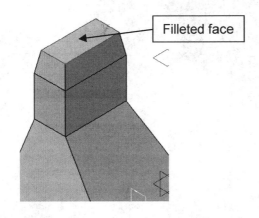

Filleted face

4) Enter the **Sketcher** on the very top surface of the part.

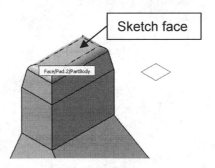

Sketch face

Face/Pad.2/PartBody

5) Draw and **Constrain** the **10 mm** diameter **Circle** shown. The circle is in the middle of the top surface.

(Problems? If your circle does not fit, go back and edit *Draft.2* and make sure that you have selected the proper neutral surface.)

6) **Exit** the *Sketcher* and **Pad** the circle to a length of **40 mm**.

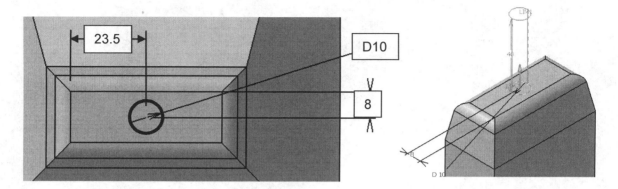

7) Create a **5 mm Edge Fillet** around the bottom edge of the pin.

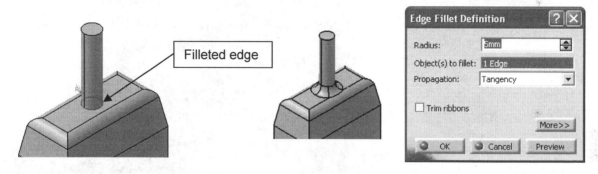

2.2 - Rolling on an Edge:

The *Rolling on an Edge* option allows you to create a fillet that has a larger radius than the distance of the sides supporting the fillet.

1) Save your Part.

2) Enter the **Sketcher** on the right face of the object between the drafted surfaces.

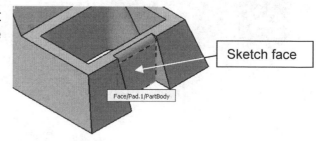

3) **3D Project Elements** the sketch face as a **Construction Element**.

4) Draw and **Constrain** the **Rectangle** as shown. Picking the top and bottom lines of the rectangle to create the 20 mm constraint will make it less likely that the rectangle will move out of position.

5) **Exit** the *Sketcher* and **Pad** 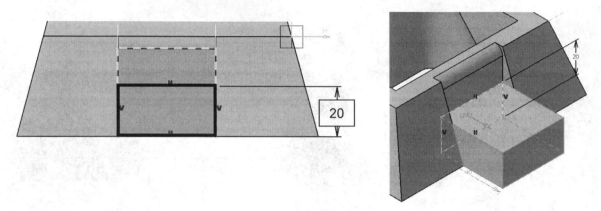 the rectangle to a length of *40 mm*.

6) Enter the **Sketcher** on the same face as before.

7) **3D Project Elements** the sketch face as a **Construction Element**.

8) Draw and **Constrain** the **Rectangle** as shown.

9) **Exit** the *Sketcher* and **Pad** 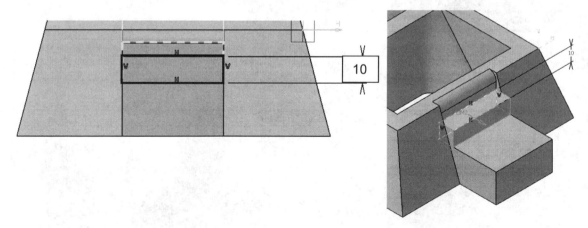 the rectangle to a length of *10 mm*.

10) Try to apply a **40 mm** radius **Edge Fillet** 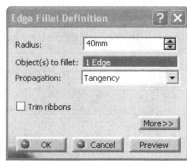 to the step. You will get an error because the step height is less than 40 mm. Select **OK** in the *Update Error* window. Select **Yes** in the *Feature Definition Error* window. When the *Edge Fillet Definition* window reappears, activate the Edge(s) to Keep: field and select the top edge of the step.

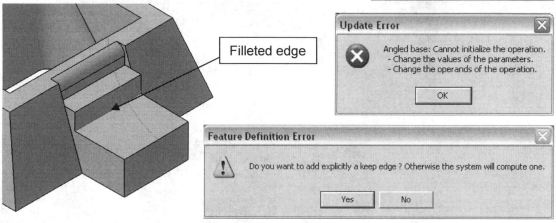

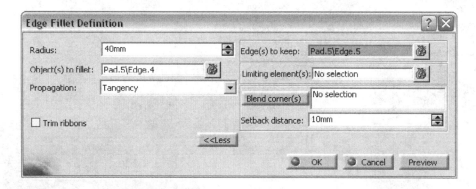

Filleted edge

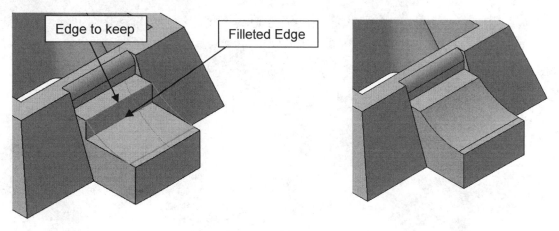

Edge to keep

Filleted Edge

11) **Save** your Part.

12) Enter the **Sketcher** on the **yz-plane**.

13) Draw and **Constrain** the **Rectangle** as shown. The reason that there are to vertical 30 mm constraints is that one constrains the height of the rectangle and one constrains the rectangle to the object.

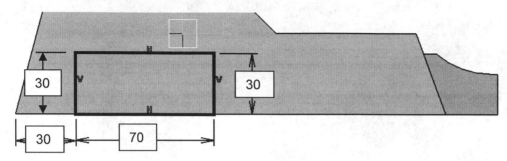

14) **Exit** the Sketcher and **Pad** the rectangle using 2 limits. The first limit has a length of **120 mm** and the second has a length of **20 mm**.

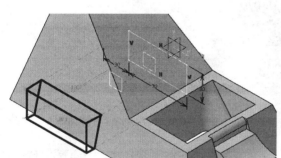
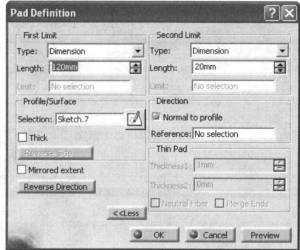

15) Apply a **10 mm Edge Fillet** to the edges shown. (Notice that the edges between the filleted edges remain sharp.)

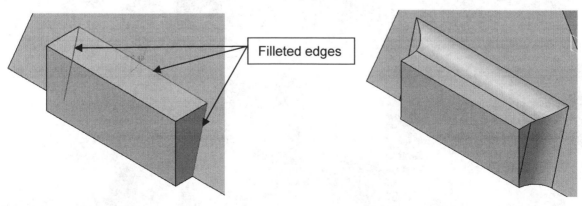

Filleted edges

16) Reenter the *Edge Fillet Definition* window by double clicking on **EdgeFillet.5**. Select the **More>>** button. Keep the 2 edges indicated. (Notice that the edges are rounded now.) Repeat for the other side.

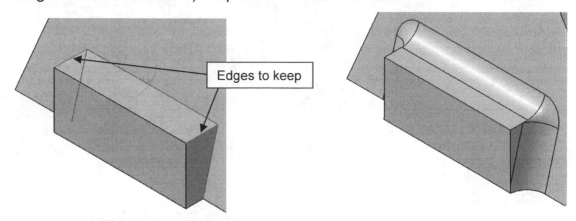

Edges to keep

2.3 - Limiting Fillets:

Limiting a fillet allows you to apply a fillet to only a portion of an edge.

1) Save your Part.

2) Apply a **10 mm** radius **Edge Fillet** to the top portion of the angled edges of the part using the **zx-plane** as a Limiting Element. Click on the orange arrow to change direction of fillet travel if necessary.

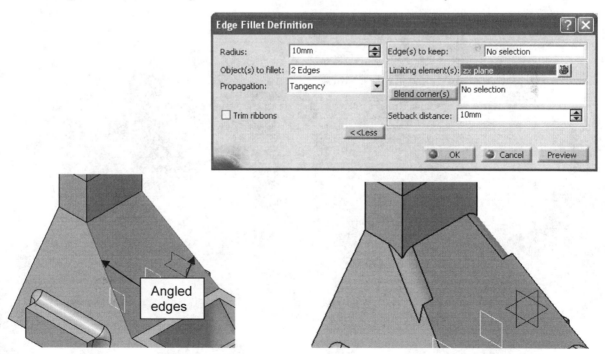

Angled edges

2.4 - Face to Face Fillets:

A *Face to Face Fillet* allows you to create a fillet between two surfaces that do not touch.

1) **Save** your Part.

2) Delete **Draft.4** and **Pocket.1** by right clicking on them in the specification tree and selecting **Delete**.

(Having Trouble? If deleting the draft or pocket causes your part to turn red, fill in the area drawing a Rectangle and Padding it 40 mm.)

3) Enter the Sketcher on the top face of the base.

4) Draw and **Constrain** the *30 mm* diameter **Circles** shown.

5) **Exit** the *Sketcher* and **Pad** the circles to a length of *20 mm*.

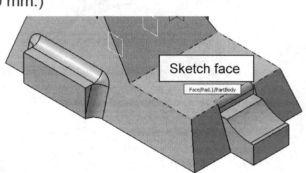

Sketch face

Face/Pad.1/PartBody

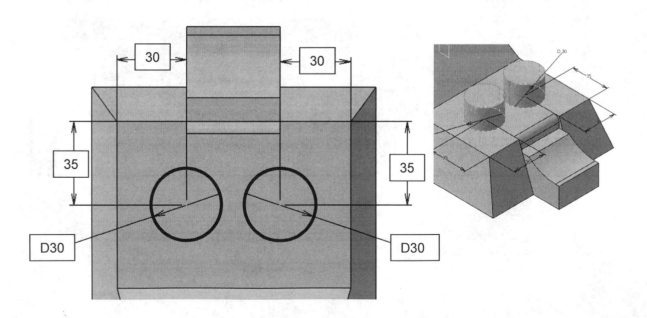

6) Put a **30° Draft Angle**  on the surfaces of the cylinders.

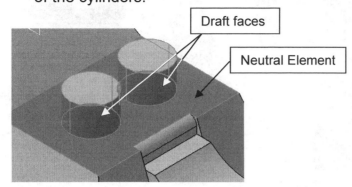

Draft faces

Neutral Element

7) Create a **15 mm Face To Face Fillet** between the two cones. (Unix users: Hold down the control key when selecting the 2 faces.) This is a complex operation. If error windows appear, close them.

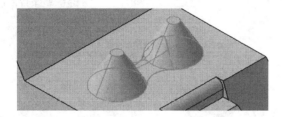

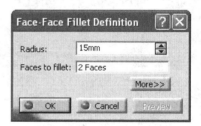

2.5 - Variable Fillets:

Variable Fillets enable you to produce a fillet with a variable radius.

1) **Save** your Part.

2) Create a **Variable Fillet** 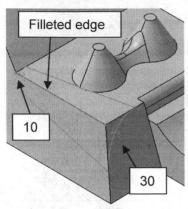 on the edge indicated that starts off at a **10 mm** radius and ends up at a **30 mm** radius using a **Linear** variation. Change the fillet radii by double clicking on the dimensions.

3) Repeat on the other side using the **Cubic** Variation.

(Note: There are Drafted filleted pads and pockets available that do both commands simultaneously.)

Section 3: Creating Chamfers

1) Create a **3 mm** x **45°** **Chamfer** 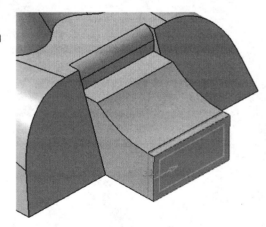 on the right face of the projection shown.

2) **Save** your Part.

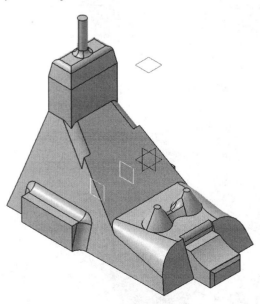

Chapter 3: PART DESIGN

Tutorial 3.6: Dress-Up Features: Shell, Thickness & Threads

Featured Topics & Commands

Prerequisite Knowledge & Commands

- The *Sketcher* workbench and associated commands
- Editing a preexisting sketch
- Pads, Pockets, Shafts and Holes
- Fillets and Drafts
- Rotating the part using the mouse or compass

Shells

The *Shell*  command is part of the *Dress-Up Features* toolbar. Shelling consists of emptying a part out while keeping a given thickness on its sides. One or more faces may be selected to be open and the remaining sides can have different thicknesses.

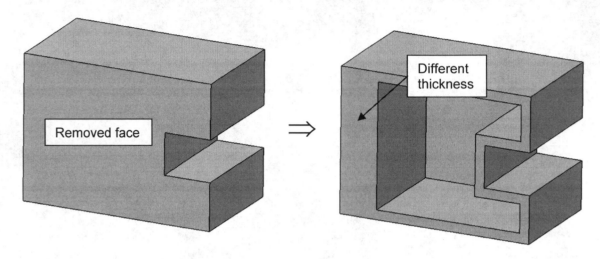

Removed face

Different thickness

$\Rightarrow$

Thickness

The *Thickness* command is part of the *Dress-Up Features* toolbar. This command allows you to add thickness to a wall.

Threads and Taps

The *Thread* command is part of the *Dress-Up Features* toolbar. The *Thread* command allows you to create threads in accordance to drawing standards. User specified threads may also be implemented. The threads cannot be viewed in the 3D space. However, they will appear on the orthographic projection of the part that is created in the *Drafting* workbench.

In the *Part Design* workbench, there is a command called

Tap – Thread Analysis . It is located in the *Analysis* toolbar. This command calls out all threads on the part. However, the call out that CATIA gives is not in accordance with ASME thread note standards.

Tutorial 3.6 Start: Part Modeled

Three different views of the part modeled in this tutorial are shown below. The part is shown as a cut away view to show the inner features. The main solid is created using the *Shaft* command and is hollowed out using the *Shell* command. *Drafts, Fillets, Threads, Thickness* and *Holes* are added to finish the part. Notice that the threads cannot be seen in the 3D view, but are represented on the section view created in the *Drafting* workbench.

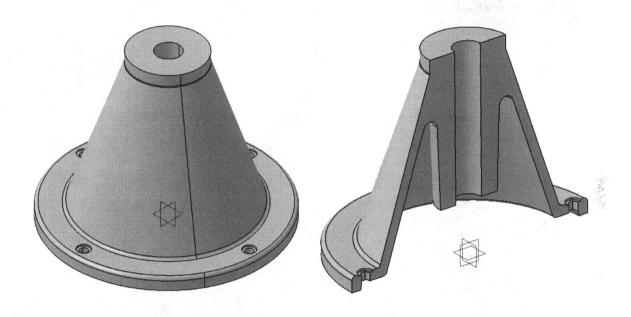

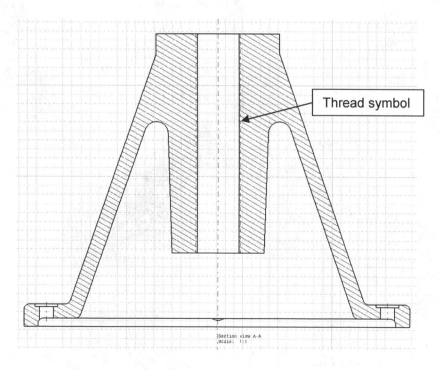

Thread symbol

Section 1: Applying a Shell.

1) Open a **New… Part** drawing and name it ***Threaded Support***.

2) Save your part as ***T3-6.CATPart***.

3) Enter the **Sketcher** on the **yz plane**.

4) Set your length units to be **inches**.

5) Draw and **Constrain** the following **Axis** and **Profile**.

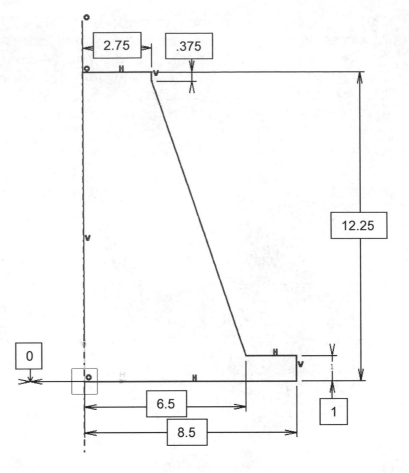

6) **Exit** the *Sketcher* and **Shaft** 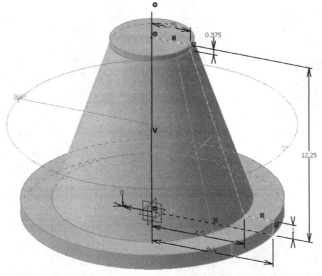 the sketch ***360 deg*** using the Sketch Axis as the Axis.

7) Pull out the *Dress-Up Features* toolbar .

8) **Shell** 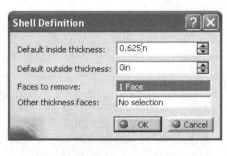 the part using a uniform thickness of *.625 in* and removing the bottom face of the part.

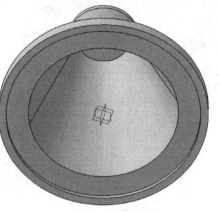

Remove this face.

9) **Edge Fillet** the edges indicated in the figure with a *0.25 in* radius. 5 edges in total.

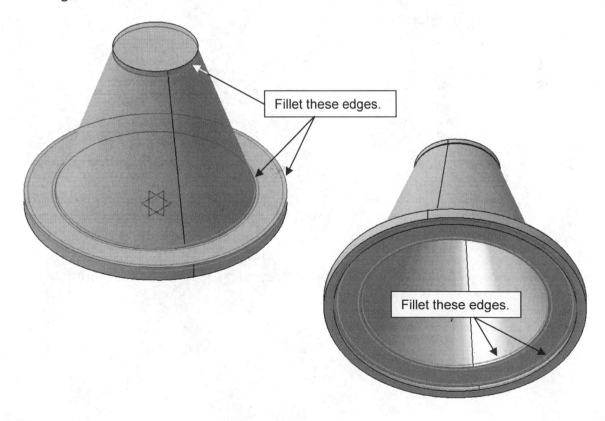

Fillet these edges.

Fillet these edges.

10) Enter the **Sketcher** on the top face of the part.

11) Draw and **Constrain** the **Circle** shown. The circle's center coincides with the origin.

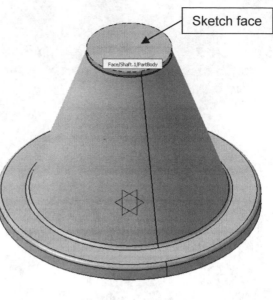

Sketch face

Face/Shaft.1/PartBody

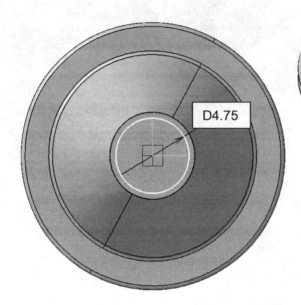

D4.75

12) **Exit** the *Sketcher* and **Pad** the sketch through the part **9 in**.

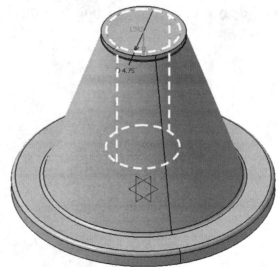

D 4.75

13) Change your view mode to **Wireframe**. This icon is located in the *View* toolbar located in the bottom toolbar area.

14) **Save** your Part.

15) Put a **Draft Angle** of *2 deg* on the padded cylinder that travels through your part using the top surface of the part as the Neutral Element. You may have to reverse the orange arrow so that the cylinder becomes smaller as it travels down.

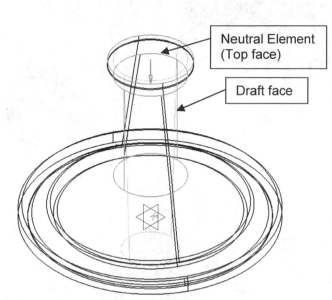

Neutral Element
(Top face)

Draft face

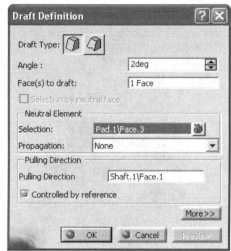

16) Change your view mode back to **Shading with Edges** 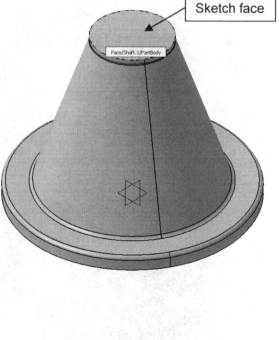.

17) Enter the **Sketcher** on the top face of the part.

Sketch face

18) Draw and **Constrain** the **Circle** shown. The circle's center coincides with the origin.

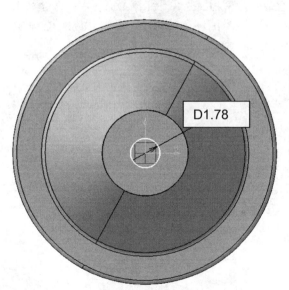

D1.78

19) **Exit** the *Sketcher* and **Pocket**  the sketch using the **Up to last** option.

20) **Edge Fillet**  the edge indicated in the figure with a *0.5 in* radius.

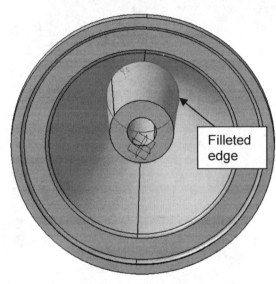

Filleted edge

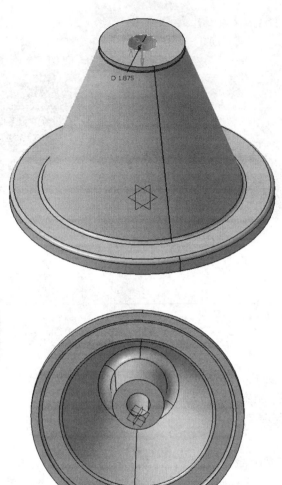

Section 2: Adding Thickness

1) Add 0.5 in of thickness to the top face of the part. Select the **Thickness** icon. Activate the `Default thickness faces:` field and select the top face of the part. Enter a thickness of *0.5 in* in the `Default thickness:` field.

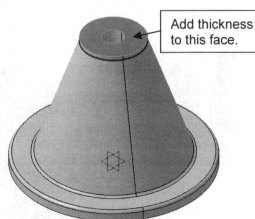

Add thickness to this face.

Thickness Definition

Default thickness:	0.5in
Default thickness faces :	1 Face
Other thickness faces :	No selection

OK   Cancel

Section 3: Creating Threads

1) Save your Part.

2) Apply a thread to the hole running down through the part. Select the **Thread** icon. Fill in the following fields;

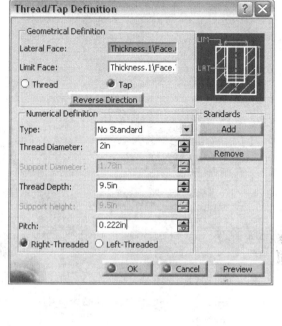

- Lateral Face: Select the circumferential surface of the hole
- Limit Face: Select the top surface of the part
- Type: **No Standard**
- Thread Diameter: *2 in*
- Thread Depth: *9.5 in*
- Pitch: *.222 in*
- Activate the **Right-Threaded** toggle.
- Notice that even though you can not see the threads, the specification tree contains the thread information.

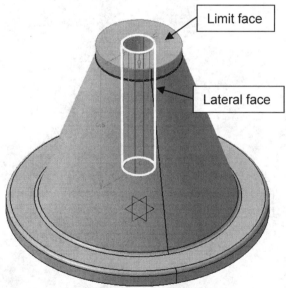

Limit face

Lateral face

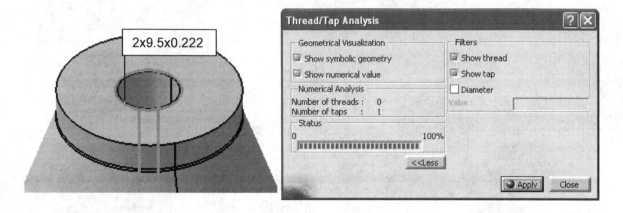

3) Select the **Thread – Tap Analysis** icon [image] in the *Analysis* toolbar. In the *Thread/Tap Analysis* window select the **More>>** button to see all the available options and then select **Apply**. Notice that a call out appears on the thread giving the thread information. Note that a more conventional thread call out for this particular thread would be **2 – 4.5 UNC – ▽9.5** where 4.5 is one over the pitch.

4) Create a Counterbored **Hole** [icon] in the bottom lip of the part with the following parameters;
   - Counterbore diameter: *1 in*
   - Counterbore depth: *0.125 in*
   - Hole diameter: *0.625 in*.

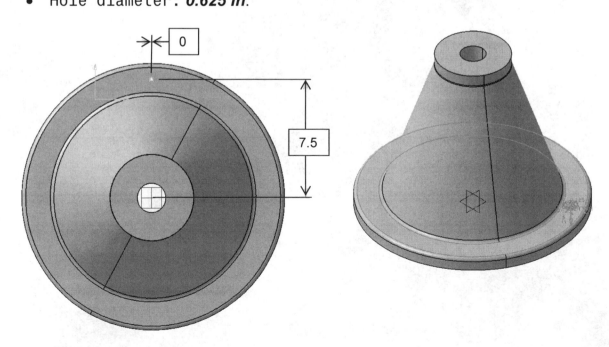

5) Create 3 more counter bored holes using a *Circular Pattern*. Select the

Circular Pattern 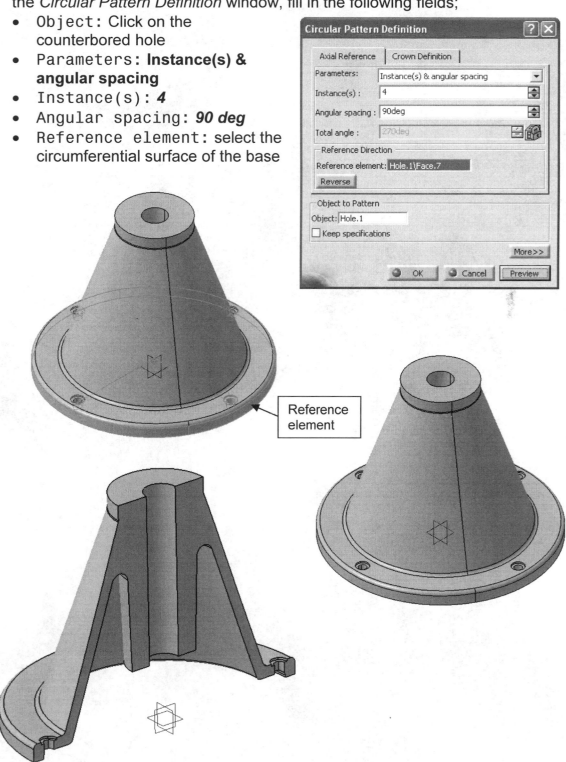 icon located in the *Transformation Features* toolbar. In the *Circular Pattern Definition* window, fill in the following fields;

- Object: Click on the counterbored hole
- Parameters: **Instance(s) & angular spacing**
- Instance(s): *4*
- Angular spacing: *90 deg*
- Reference element: select the circumferential surface of the base

Circular Pattern Definition

| Axial Reference | Crown Definition |

Parameters: Instance(s) & angular spacing

Instance(s): 4

Angular spacing: 90deg

Total angle: 270deg

Reference Direction

Reference element: Hole.1\Face.7

Reverse

Object to Pattern

Object: Hole.1

☐ Keep specifications

More>>

OK   Cancel   Preview

Reference element

6) **Save** your Part.

NOTES:

Chapter 3:
PART DESIGN

Tutorial 3.7: Transformation Features

Featured Topics & Commands

Prerequisite Knowledge & Commands

- The *Sketcher* workbench and associated commands
- Editing a preexisting sketch
- Pads and Pockets
- Rotating the part using the mouse or compass

The Transformation Features toolbar

The *Transformation Features* toolbar contains commands that transform, move or create a pattern or array of a feature. Reading from left to right, the commands located in the *Transformation Features* toolbar are

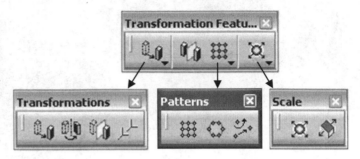

- The Transformation toolbar:
  The *Transformation* toolbar contains commands that allow you to move, rotate and create a mirror image of a part.
- Mirror: The *Mirror* command creates a duplicate mirror image of a part.
- The Patterns toolbar: The *Pattern* toolbar allows you to create arrays of an existing feature.
- The Scale toolbar: The *Scale* toolbar contains commands that may be used to scale a part. The *Scale* command changes the size of a part relative to a selected plane or point. Scaling can only occur in one direction at a time. The *Affinity* command transforms the absolute size by multiplying a scale ratio.

The Transformation toolbar

A transformation is the ability to move a body either by translating it along an axis, rotating it around an axis or moving it symmetrically about a plane. A body cannot be duplicated during transformation. Reading from left to right, the commands located in the *Transformation* toolbar are

- Translation: *Translation* moves the part in a selected linear direction.
- Rotation: *Rotation* rotates a part around a selected line or axis.
- Symmetry: *Symmetry* replaces the part with its mirror image.
- AxisToAxis: *AxisToAxis* transform a geometry's position relative to one axis system into a new axis system.

When transforming a part, CATIA will prompt you with a *Question* window to make sure it is okay to violate the constraints that were imposed in the *Sketcher*. If you are sure you want to continue with the transformation, click <u>Y</u>es.

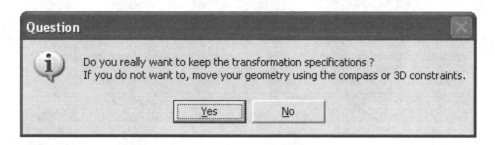

The Patterns toolbar

Patterns allow you to create arrays of an existing feature. You can create rectangular or polar arrays. The positions of the feature may also be manually defined. Patterned features are connected or associative. Therefore, if the original feature is changed all the patterned features will change. You can break the connection between the features by exploding the pattern. To explode the pattern, right click on the pattern feature in the specification tree and select *Patter<u>n</u> object - <u>E</u>xplode...* Reading from left to right, the commands located in the *Patterns* toolbar are

- <u>Rectangular Pattern:</u> This command enables you to create rectangular arrays of a feature with a specified number of rows and columns.
- <u>Circular Pattern:</u> This command enables you to create polar arrays of a feature.
- <u>User Pattern:</u> This command allows you to create several instances of a feature in predefined locations. The locations are defined by the user in a sketch which usually consists of several points.

Tutorial 3.7 Start: Part Modeled

The part modeled in this tutorial is a simplified representation of a bicycle chain ring. It contains some features that do not exist on a chain ring, but are created to illustrate a particular command. Ignoring the rectangular holes in the right arm, notice the parts symmetry and repeated features.

Section 1: Creating circular patterns.

1) Open a **New… Part** drawing, name it **Sprocket**.

2) Save your part as **T3-7.CATPart**.

3) Enter the **Sketcher** on the **yz plane**.

4) Set your length units to be **inches**.

5) Draw and constrain the following sketch. Notice that the center of the arc is coincident with the origin.

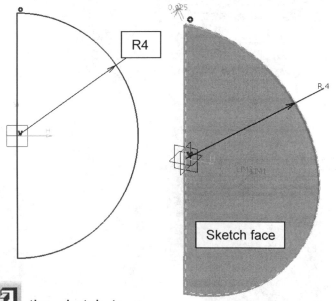

6) **Exit** the Sketcher and **Pad** [icon] the sketch to a length of **0.125 in**.

7) Enter the **Sketcher** on the front face of the Sprocket.

8) Draw and constrain the following **Construction Element** sketch.

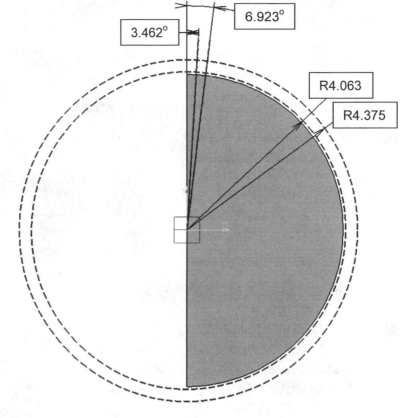

9) **3D Project Elements** the face of the part on to the sketch plane as a **Construction Element**.

10) Using the construction lines as a guide, draw and constrain the sketch shown. The sides of the tooth are constructed using the **Profile** command and the *Tangent arc* option. The top of the tooth is created using the *Three Point Arc* option.

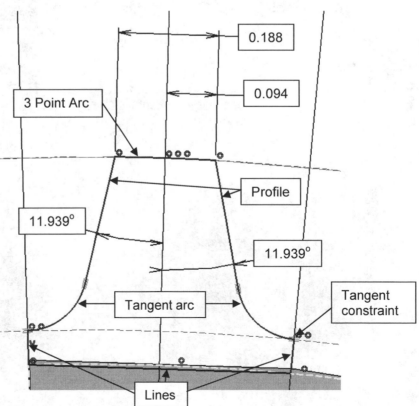

11)**Exit** the Sketcher and **Pad**  the sketch to a length of *.125 in*. Reverse the direction if necessary.

12)Pull out the *Transformation Features* toolbar

13)**Save** your Part.

14)Create a circular pattern of the chain ring tooth. Select the **Circular Pattern** icon. In the *Circular Pattern Definition* window, fill in the following fields;

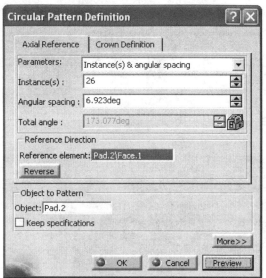

- Parameters: **Instance(s) & angular spacing**
- Instance(s): *26*
- Angular spacing: *6.923deg*
- Reference element: Select the top of the tooth (You may have to zoom in to do this.)
- Object: select **Pad.2** (the pad used to create the tooth) in the specification tree. Reverse the direction if necessary.

Reference element for pattern.

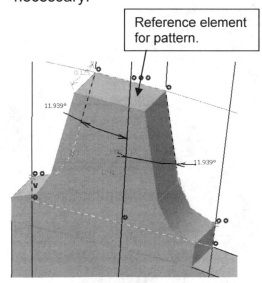

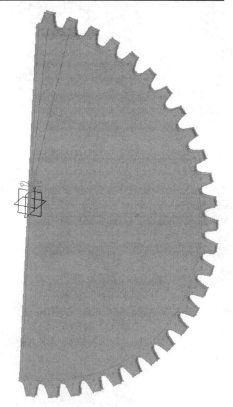

Section 2: Mirroring a part.

1) Mirror your half chain ring to create a full one. Select the

   Mirror 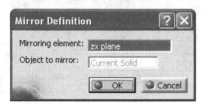 icon. Select the **zx plane** as your Mirroring element:. Read the prompt line for help.

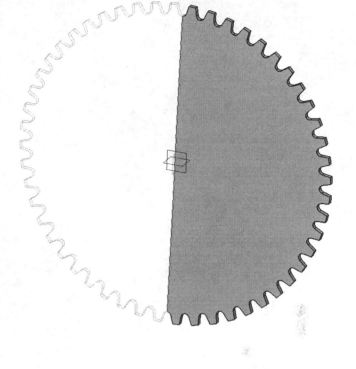

(Problems? If you get an error when trying to mirror the object, check to make sure that the bottom line of Sketch.2 (the tooth) goes completely below Pad.1.)

2) Enter the **Sketcher** on the **zx plane**.

3) Draw and **Constrain** the following **Rectangle**.

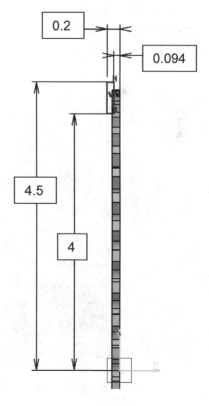

4) **Exit** the *Sketcher* and create a **Groove**  using the sketch that travels **360 deg** using the **HDirection** as the Axis.

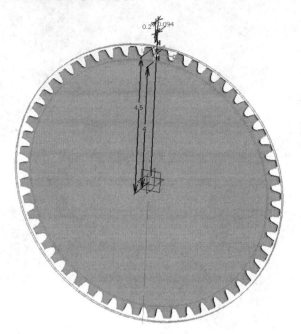

Groove Definition [?][X]

Limits

First angle: 360deg

Second angle: 0deg

Profile/Surface

Selection: Sketch.4

☐ Thick Profile

Reverse Side

Axis

Selection: HDirection

Reverse Direction

More>>

OK   Cancel   Preview

5) Enter the **Sketcher** on the front face of the part. Draw and constrain the following sketch. The length of the lines have to be long enough to go past the two circles, but the length beyond that doesn't matter.

6) Use the **Quick Trim** command to create a slot from the lines and circles previously draw.

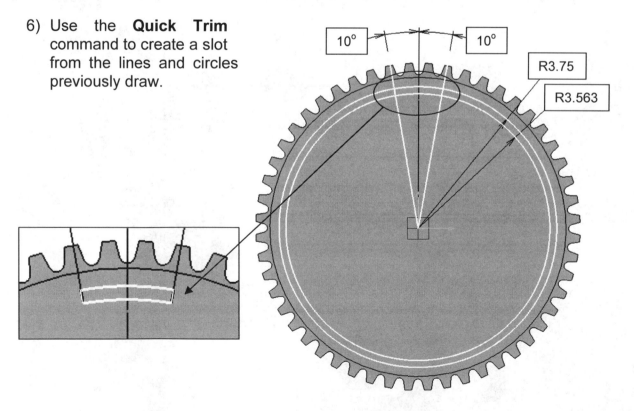

10°   10°

R3.75

R3.563

7) Apply *0.09 in* radii **Corners** to the four sharp corners of the slot.

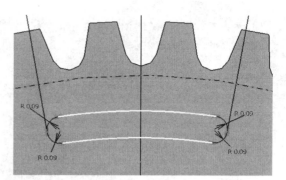

8) **Exit** the Sketcher and **Pocket** 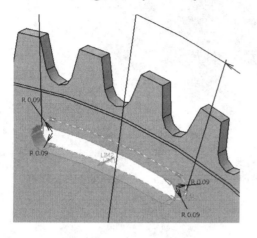 the sketch using the option **Up to last**.

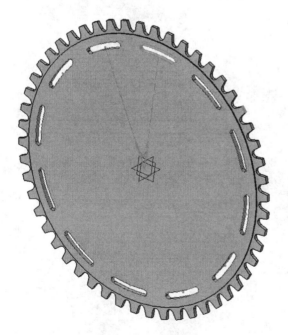

9) Save your part.

10) Create a **Circular Pattern** of the slot creating *12* instances spaced *30 deg* apart. Use the top of the tooth (as before) as the reference element.

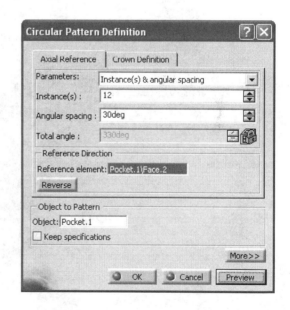

11) Enter the **Sketcher** on the front face of the part. Draw and constrain the following sketch.

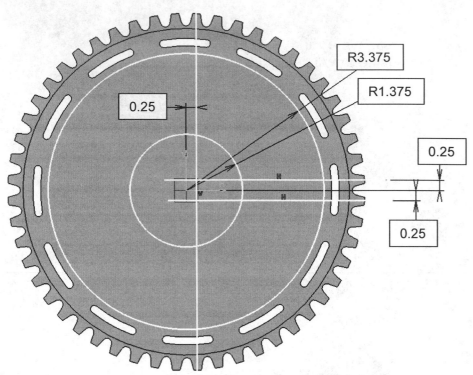

12) Use the **Quick Trim** command to create the two shapes shown from the lines and circles previously draw.

13) **Exit** the Sketcher and **Pocket** 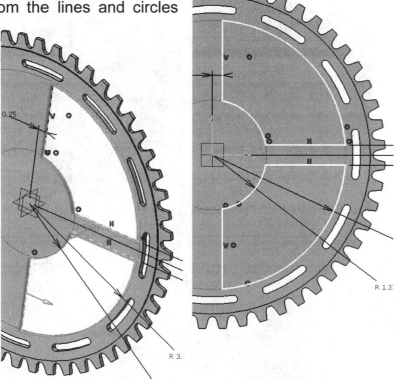 the sketch using the option **Up to last**.

14) Mirror Pocket.2. Select the
Mirror  icon and then select
Pocket.2 in the specification tree.
Select the **zx plane** as your
Mirroring element:. Read the
prompt line for help.

15) Save your part.

16) Enter the **Sketcher** on the front
face of the part. Draw and
constrain the following sketch.

17) **Exit** the Sketcher and **Pocket**
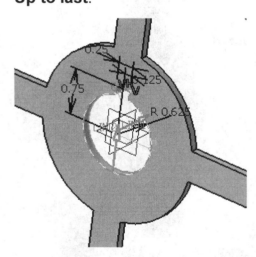 the sketch using the option
Up to last.

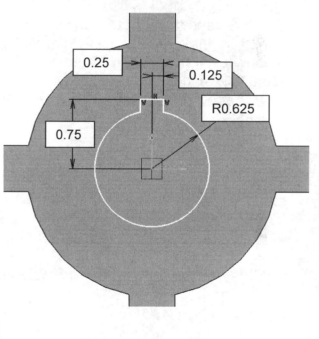

Section 3: Creating rectangular patterns.

1) Save your part.

2) Enter the **Sketcher** on the front face of the part.   Draw and constrain the following sketch.

3) **Exit** the Sketcher and **Pocket** 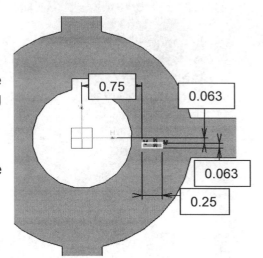 the sketch using the option **Up to last**.

4) Create a rectangular pattern of the rectangular slot that produces 2 rows 5 slots along the

arm.   Select the **Rectangular Pattern** icon.   In the *Rectangular Pattern Definition* window select the **First Direction** tab and fill in the following fields;

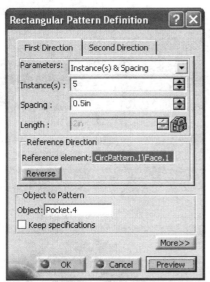

- Parameters: **Instance(s) & Spacing**
- Instance(s): *5*
- Spacing: *0.5 in*
- Reference element: select the front face of the part
- Object: **Pocket.4** (The pocket associated with the rectangular pocket).

Select the **Second Direction** tab and fill in the following fields;

- Parameters: **Instance(s) & Spacing**
- Instance(s): *2*
- Spacing: *0.125 in*
- Reference element: select the front face of the part
- Object: **Pocket.4**

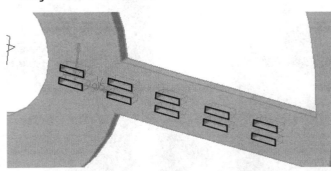

5) Edit Sketch.7 (the sketch associated with the rectangular slot) and change the height of the rectangle from 0.063 in to **0.094 in**.

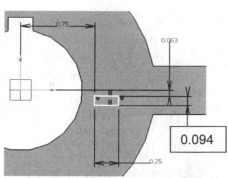

6) **Exit** the Sketcher. Notice that all the rectangular slots in the array update based on the original.

7) Right click on RectPattern.1 in the specification tree and select **RectPatter_n.1 object – E_xplode...** We have just broken the connection between the array elements. Notice that the RectPattern.1 feature is now broken up in to 10 individual pocket features.

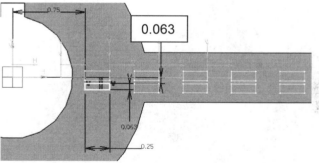

8) Edit Sketch.7 and change the height of the rectangle back to **0.063 in**.

9) **Exit** the Sketcher. Notice that only the Sketch.7 updates. The other slots are unaffected.

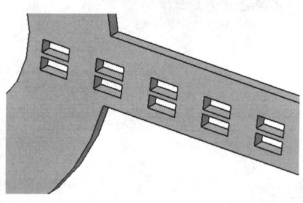

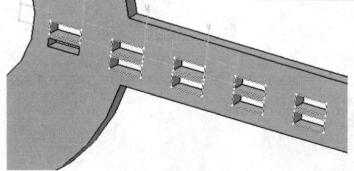

Section 4: Transformations.

When performing the transformations in this section, click on *Yes* in the *Question* window when prompted to (If one appears).

1) Save your part.

2) Scale the thickness of the Sprocket by a factor of 2.

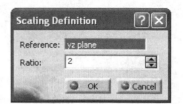

Select the **Scale** 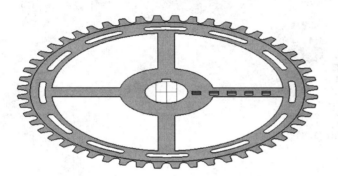 icon and set the reference to be the **yz plane** and the ratio to be **2**.

3) Scale the part by *0.5* using the **xy plane** as a reference.

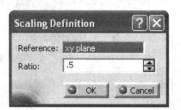

4) **Rotate** the part by *45 deg* using the **X Axis** as the rotation axis.

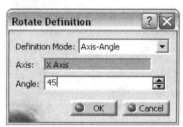

5) Select the **Undo with history** icon. Undo all the transformations performed in this section.

NOTES:

NOTES:

Chapter 3:
PART DESIGN

Tutorial 3.8: Boolean Operations

Featured Topics & Commands

Prerequisite Knowledge & Commands

- The *Sketcher* workbench and associated commands
- Editing a preexisting sketch
- Pads and Shafts
- Creating reference planes
- The *Transformation* toolbar
- Rotating the part using the mouse or compass

The Boolean Operations toolbar

Boolean operations allow you to create parts from individual solids. The commands located in the *Boolean Operations* toolbar, reading from left to right are

- Assemble:
  - Add: Adding a body to another one means uniting them and creating a single body.
  - Remove: This command removes the intersecting volume of one body from another. The first body is then deleted.
  - Intersect: This command removes everything except the volume that is created when two bodies intersect.
- Union Trim: This command allows you to add to bodies together while specifying which parts to keep and which parts will be trimmed.
- Remove Lump: This command allows you to reshape a body by removing material.

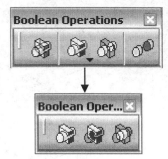

New part bodies

When you create a solid distant from your original pad, it is still considered to be the same part. If you move one, the other moves. In the specification tree, these two pads are under the same PartBody. In some instances, it is desirable to create a solid that can be moved and transformed independently of the other solids that already exist. In order to accomplish this, a new body has to be inserted. At the top pull down menu you would select *Insert – Body*. Now, any new solid that is created will be independent of the original PartBody.

When your part drawing includes several bodies, you can then associate these bodies in different ways (see the Assemble, Add, Union Trim, Intersect, Remove and Remove Lump commands) to obtain the final shape of the part.

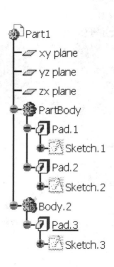

Local Axis

The *Axis System* command is used to create a coordinate system away from the three original coordinate planes. The origin and axis directions may be defined. This new coordinate system may also be rotated.

Tutorial 3.8 Start: Part Modeled

The part modeled in this tutorial is shown below. It is created using the three main Boolean Operations (Add, Intersect, and Remove). New part bodies are inserted frequently to allow independent transformations.

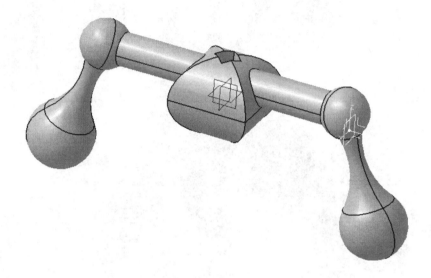

Section 1: Inserting a new part body.

1) Open a **New... Part** drawing.   If prompted, name it **Handle**.

2) Save your part as **T3-8.CATPart**.

3) Enter the **Sketcher** on the **zx plane**.

4) Set your length units to be millimeters (**Tools** – **Options...** – **Parameters and Measures**).

5) Draw and **Constrain** the **Circle** shown. The center of the circle is located at the origin.

6) **Exit** the *Sketcher* and **Pad** the sketch **30 mm** in <u>both directions</u>.

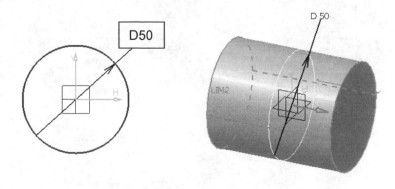

7) At the top pull down menu, select **Insert – Body**.

8) Enter the **Sketcher** on the **yz plane**. Draw and **Constrain** the **Circle** shown. The center of the circle is located at the origin.

9) **Exit** the *Sketcher* and **Pad** the sketch **30 mm** in both directions. Your specification tree should look like the one shown.

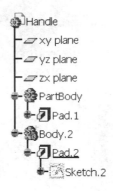

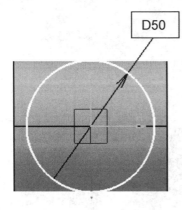

D50

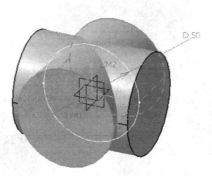

Section 2: Boolean operations.

1) Activate your Boolean operations toolbar if not already active. At the top pull down menu, select **View – Toolbars – Boolean Operations**.

2) Deselect all.

3) Select the **Intersect** icon and select the two cylindrical part bodies either in the tree or by selecting the bodies themselves. Select **OK** in the Intersect window.

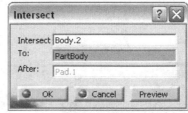

4) At the top pull down menu, select **Insert – Body**.

5) Enter the **Sketcher** on the **zx plane**. Draw and **Constrain** the **Circle** shown. The center of the circle is located at the origin.

6) **Exit** the *Sketcher* and **Pad** the sketch *70 mm* in both directions. Your specification tree should look like the one shown.

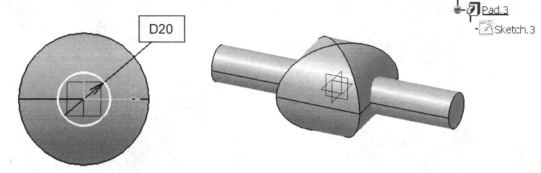

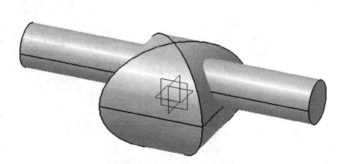

7) Select the **Translate** 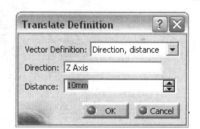 icon. Select **Yes** in the *Question* window. Translate Body.3 *10 mm* in the **Z Axis** direction.

8) Deselect all.

9) Select the **Add**  icon and add **Body.3** to the original **PartBody**.

10) Save your part.

11) At the top pull down menu, select **Insert** – **Body**.

12) Enter the **Sketcher** on the **xy plane**. Draw and **Constrain** the **Oriented Rectangle** as shown.

(Having Trouble? If you are having trouble constraining the 6 mm dimensions, constraint the points of the rectangle with the H and V axis instead.)

13) **Exit** the *Sketcher* and **Pad** the sketch **30 mm** in both directions.

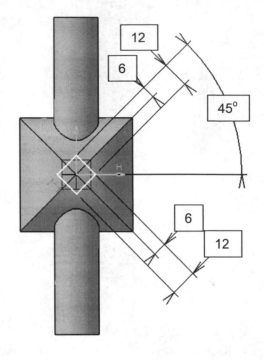

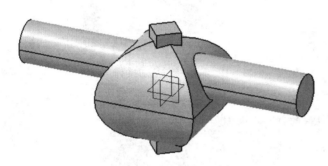

14) **Remove** 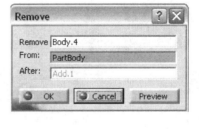 Body.4 from the original PartBody.

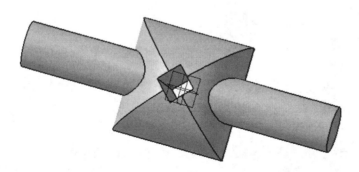

15) Save your part.

16) At the top pull down menu, select **Insert** – **Body**.

17) Enter the **Sketcher** on the end of the shaft.

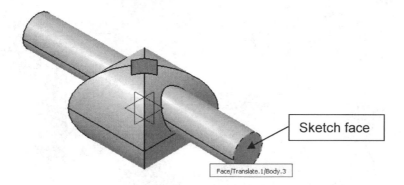

Sketch face

Face/Translate.1/Body.3

18) **3D Project Elements** the end of the shaft as a **Construction Element**.

19) Draw and **Constrain** the **Axis** and **Arc** shown.

20) **Exit** the *Sketcher* and **Shaft** the sketch ***360°***.

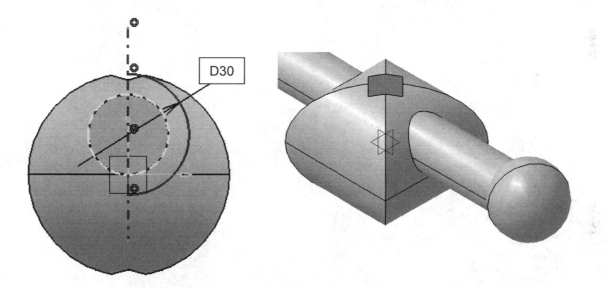

D30

21) **Translate** 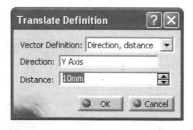 Body.5 (the sphere) ***10 mm*** in the ***Y Axis*** direction.

22) Insert a new Body.

Translate Definition

Vector Definition: Direction, distance
Direction: Y Axis
Distance: 10mm

OK   Cancel

23) Enter the **Sketcher** on the **yz plane**. Draw and **Constrain** the **Axis** and **Profile** shown.

24) **Exit** the *Sketcher* and **Shaft** the sketch *360°*.

25) **Add** Body.5 to Body.6.

26) Save your part.

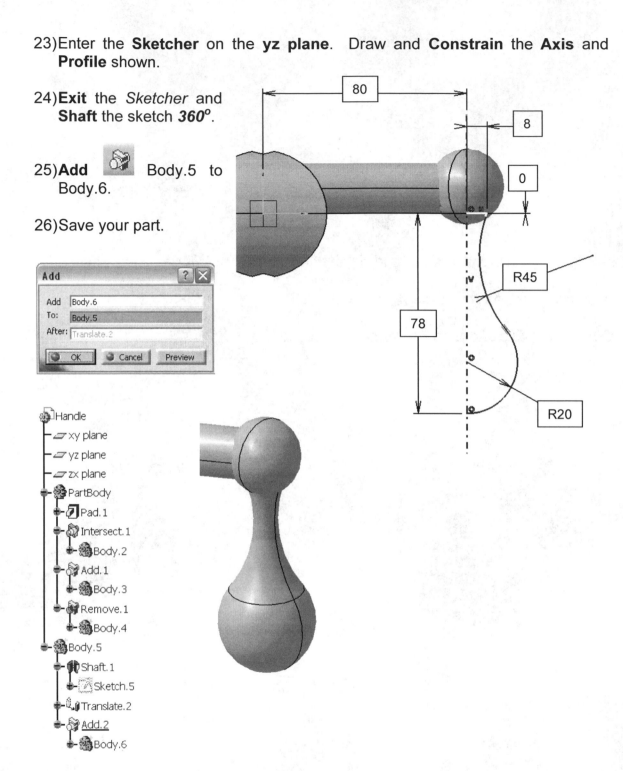

Section 3: Local axes.

1) Create a new **Plane** that is offset **80 mm** from the **zx plane**.

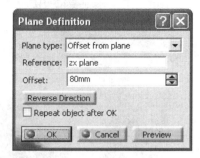

2) In the *Tools* toolbar, select the **Axis System** icon. In the *Axis System Definition* window, fill in the following fields;
   - Y axis: select the new plane (Plane.1)
   - Z axis: select the **xy plane**. (Note: The new axes will be perpendicular to the plane selected to define them.)

3) **Rotate** 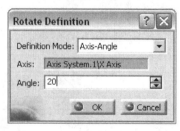 the handle grip (Body.5+Body.6) by **20°** using the newly created **X axis** as the rotation axis.

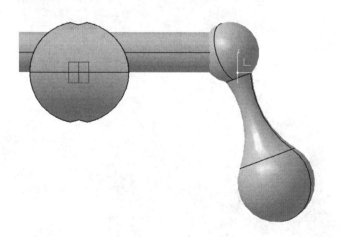

4) **Mirror** 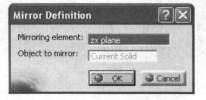 the handle using the **zx plane** as the mirroring element.

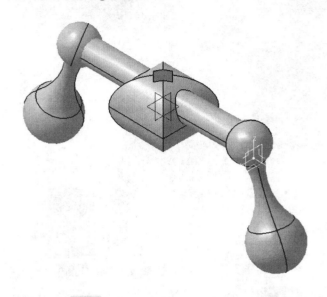

5) **Add** the handle (Body.5) to the main PartBody.

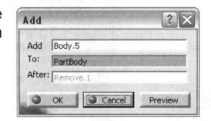

6) At completion, the specification tree should look similar to that shown on the right.

7) **Save** your part.

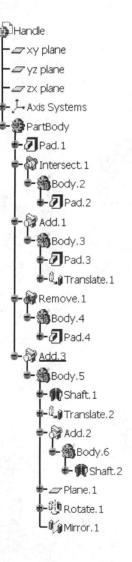

Chapter 3: PART DESIGN

Tutorial 3.9: Material Properties

Featured Topics & Commands

Prerequisite Knowledge & Commands

- The *Sketcher* workbench and associated commands
- Editing a preexisting sketch
- Pads
- Rotating the part using the mouse or compass

Applying and modifying material properties.

A specific material may be assigned to your part by using the *Apply Material* icon. There are many predefined materials available to choose from including metals, plastics, and woods. After a material has been assigned to the part, its properties may be modified to suit your particular needs. This is done by right clicking on the material in the specification tree and selecting *Properties*. Once the correct material properties have been assigned to your part, the part

properties may be obtained by selecting the *Measure Inertia* icon. This is located in the *Measure* toolbar. This command gives you information on the part's volume, area, mass, and moments of inertia. It is also used to modify how the part looks or is rendered, and the type of section line symbol that will be used in any section views that are created.

Tutorial 3.9 Start: Part Modeled

The part modeled in this tutorial is simple. It is only used as a base for learning about applying and modifying materials and analyzing part properties.

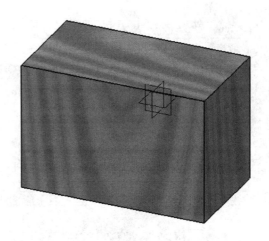

Section 1: Applying Material.

1) Open a **New… Part** drawing. If prompted, name it **Block**.

2) **Save** your part as **T3-9.CATPart**.

3) Enter the **Sketcher** on the **yz plane**.

4) Before you start to draw make sure your length units are set to millimeters. (**Tools – Options... – Parameters and Measures**)

5) Draw and **Constrain** the **Rectangle** shown.

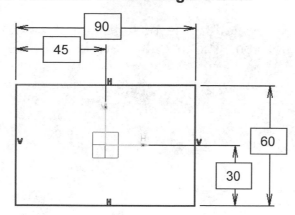

6) **Exit** the *Sketcher* and **Pad** the sketch to a length of *50 mm*.

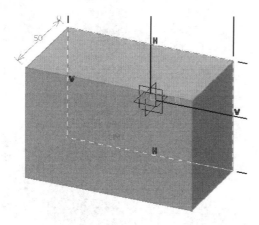

7) At the top pull down menu, select **View** – **Render Style** – **Customize View**. In the *Custom View Modes* window, activate the **Materials** toggle.

8) Select **PartBody** in the specification tree and then select the **Apply Materials** icon. This is usually located in the bottom toolbar area. Select the **Metal** tab and then the **Steel** icon. Then **Apply Material** and select **OK**.

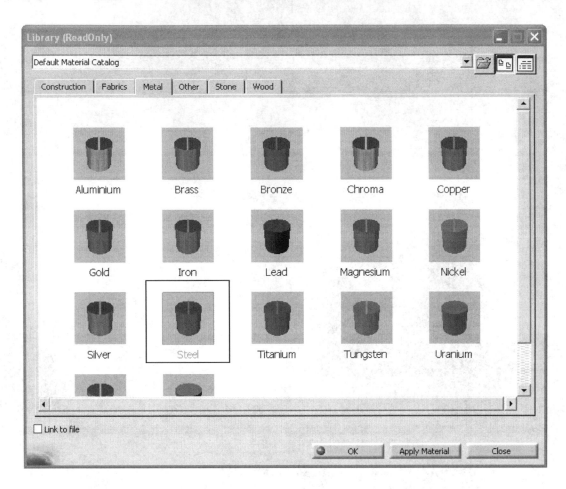

Section 2: Part properties.

1) Select **PartBody** in the specification tree and then the **Measure Inertia** icon. Usually located in the bottom toolbar area. Notice that the *Measure Inertia* window gives the part's volume, area, mass, center of gravity, and moments of inertia about the center of gravity. This information may be exported to a text file. For a simple part this is trivial information and

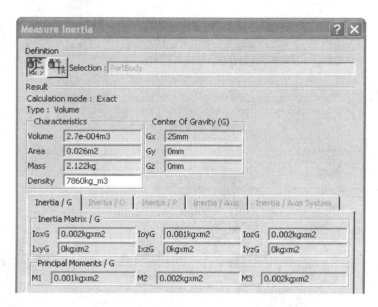

can be calculated by hand. However, for more complex parts this information is not easily calculated by hand.

2) In the *Measure Inertia* window, select the **Customize...** icon. In the *Measure Inertia Customize* window, activate the **Inertia matirix/O** toggle and select **OK**. Now the moments of inertia about the origin are calculated.

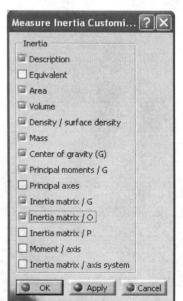

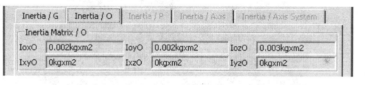

(Note: Properties for a face may also be obtained by selecting only the face and then the *Measure Inertia* icon.)

Section 3: Modifying material properties.

1) In the specification tree, right click on **Steel** and select **Properties**.

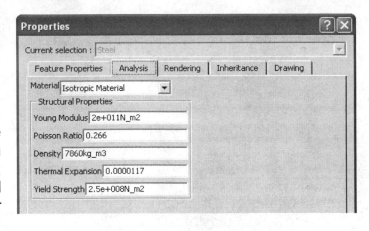

2) In the *Properties* window, select the **Analysis** tab. Notice that you can change the material type between Isotropic and Orthotropic, and the structural properties to suit your specific steel type.

3) Select the **Drawing** tab. This is where you can set your section line (hatching) style if a section view is a planned part of your orthographic projection.

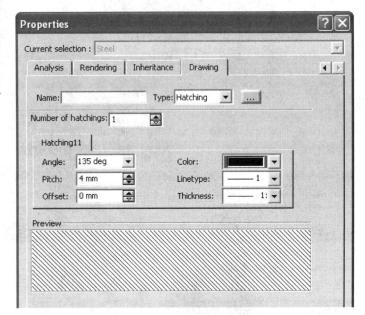

4) In your specification tree, right click on **Steel** and select **Delete**.

Section 4: Rendering

1) Select **PartBody** in the specification tree and then select the **Apply Materials**
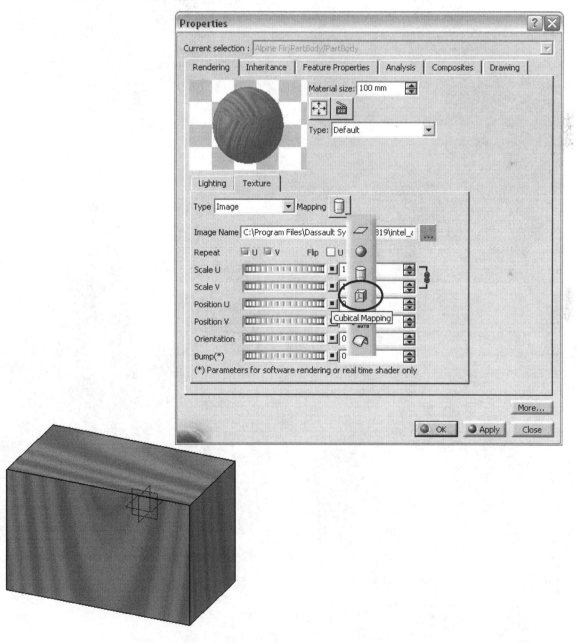 icon. This is usually located in the bottom toolbar area. Select the **Wood** tab and then the **Alpine Fir** icon. Then **Apply Material** and select **OK**.

2) In the specification tree, right click on **Alpine Fir** and select **Properties**.

3) In the *Properties* window, select the **Rendering** tab. Expand the **Change Mapping type and Preview** icon in the **Texture** tab and select **Cubical Mapping**. Change the Material size: to *100 mm* and select **Apply**. Notice the difference. Select **OK**.

<u>NOTES:</u>

Chapter 3: Exercises

Exercise 3.1: This exercise can be performed after completing tutorial 3.1. Model the following parts.

a) Use at least one *Multiple Pad* when modeling this part.

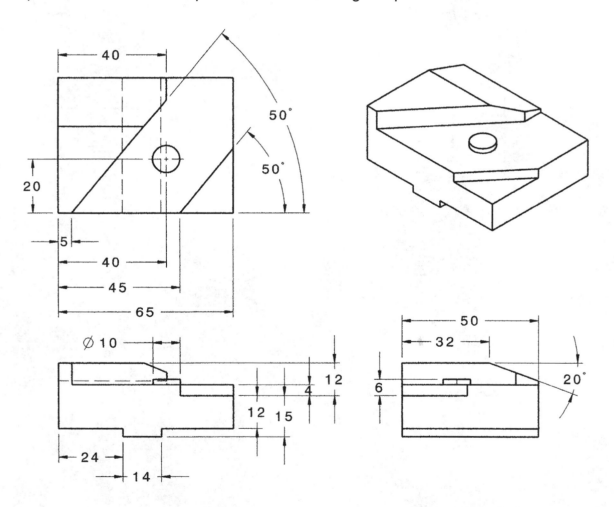

b) Model this part using only one sketch and one *Multi-Pad* command.

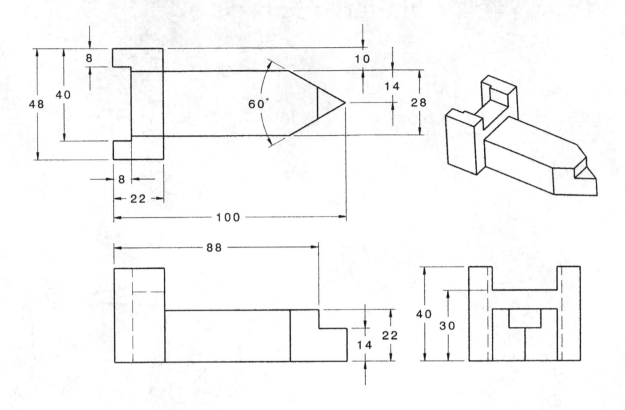

Exercise 3.2: This exercise can be performed after completing tutorial 3.2. Model the following parts applying the appropriate constraints. Use the *Hole* command to create any holes, counterbores or threads.

a)

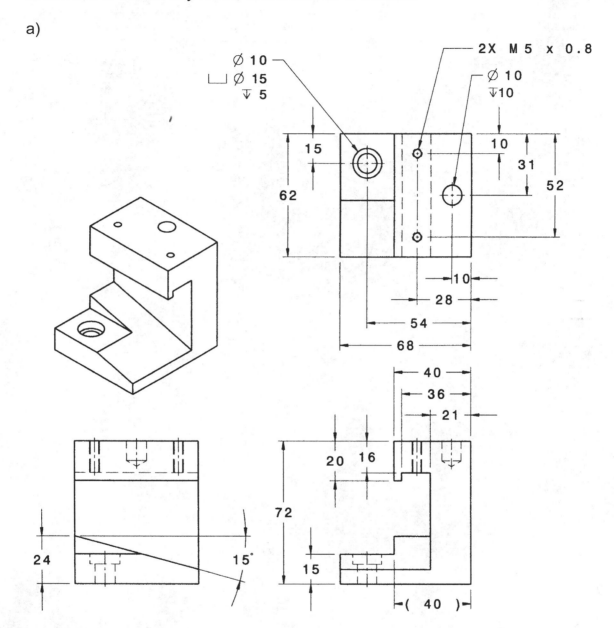

b)

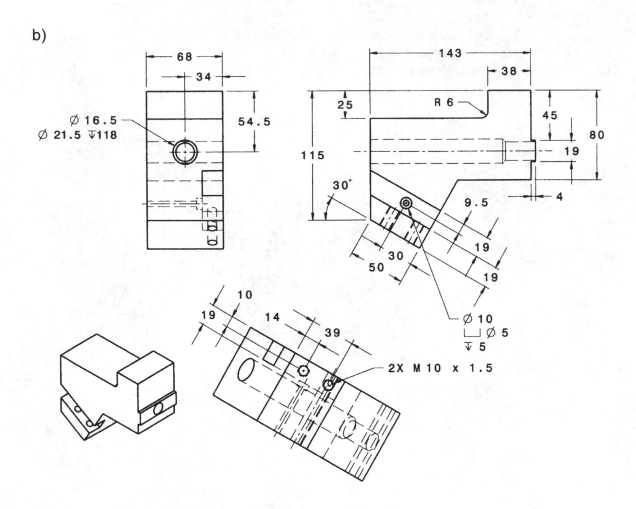

Ø 16.5
Ø 21.5 ∇118

68
34
54.5

143
38
25
R 6
45
80
19
115
30°
9.5
19
50
30
4
19

10
19
14
39

Ø 10
⌴ Ø 5
∇ 5

2X M 10 x 1.5

Exercise 3.3: This exercise can be performed after completing tutorial 3.3. Model the following parts applying the appropriate constraints. Use the *Shaft* command to create the main solid.

a)

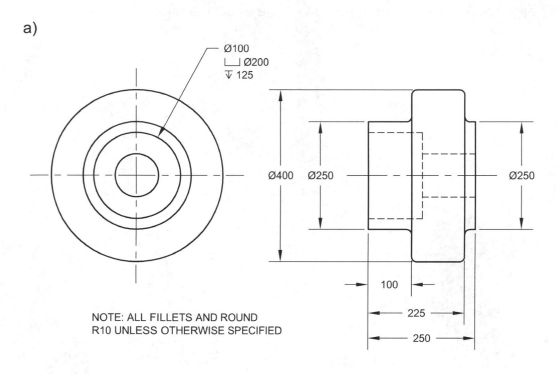

Ø100
⌴ Ø200
▽ 125

Ø400   Ø250

Ø250

100

225

250

NOTE: ALL FILLETS AND ROUND
R10 UNLESS OTHERWISE SPECIFIED

b)

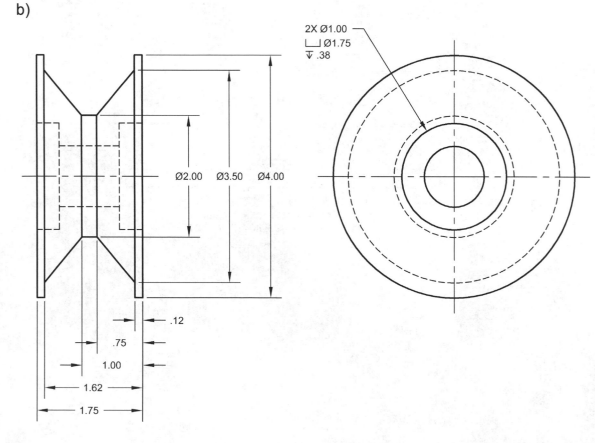

2X Ø1.00
⌴ Ø1.75
▽ .38

Ø2.00   Ø3.50   Ø4.00

.12

.75

1.00

1.62

1.75

Exercise 3.4: This exercise can be performed after completing tutorial 3.4 and 3.5. Sketch the following parts applying the appropriate constraints. Apply *Edge Fillets* where necessary and at least on *Stiffener*. If you encounter problems applying an edge fillet, apply that one first before apply the others.

a)

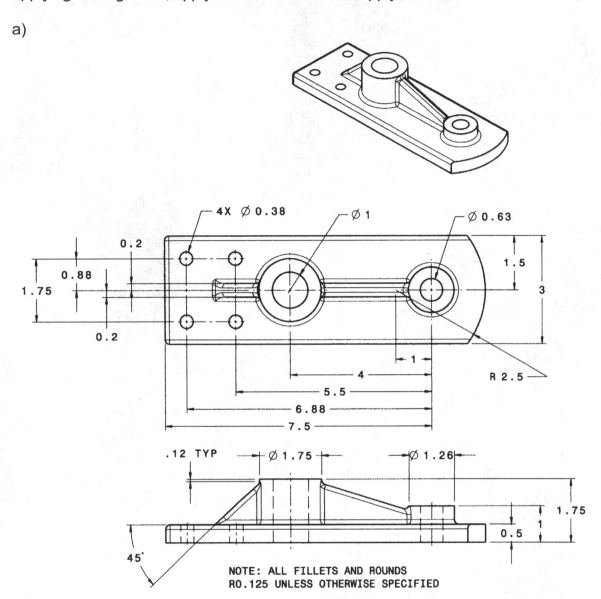

NOTE: ALL FILLETS AND ROUNDS
R0.125 UNLESS OTHERWISE SPECIFIED

b)

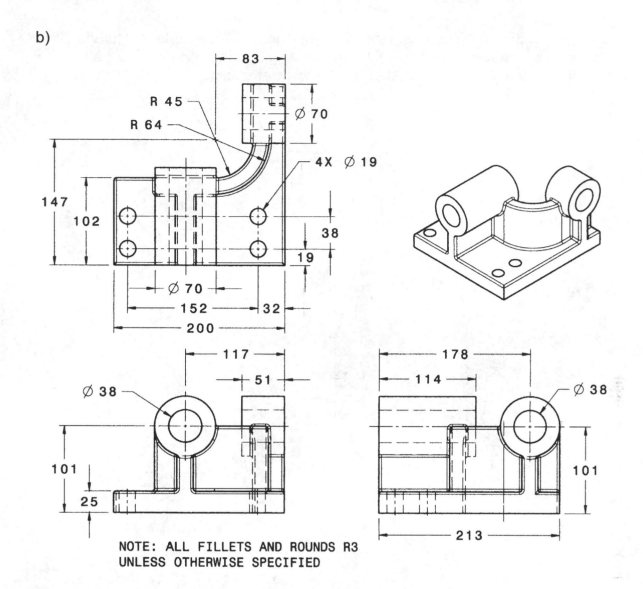

NOTE: ALL FILLETS AND ROUNDS R3
UNLESS OTHERWISE SPECIFIED

Exercise 3.5: This exercise can be performed after completing tutorial 3.7. Model the following parts applying the appropriate constraints. Use the appropriate transformations in the *Part Design* workbench.

a)

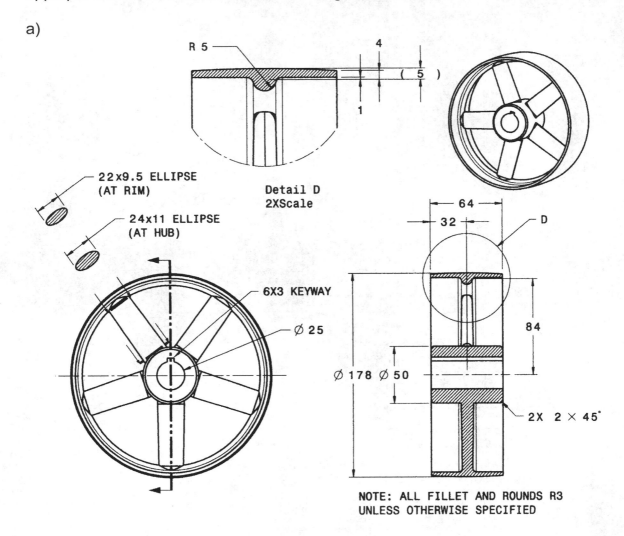

R 5

4

(5)

1

22x9.5 ELLIPSE
(AT RIM)

Detail D
2XScale

24x11 ELLIPSE
(AT HUB)

6X3 KEYWAY

Ø 25

64

32

D

84

Ø 178 Ø 50

2X 2 × 45°

NOTE: ALL FILLET AND ROUNDS R3
UNLESS OTHERWISE SPECIFIED

b)

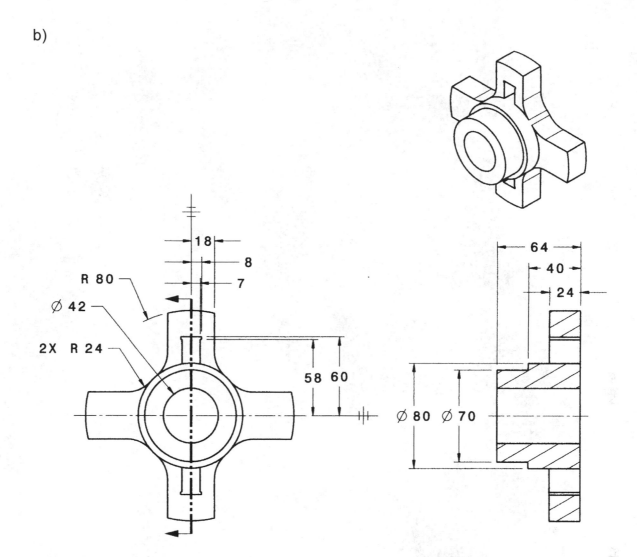

NOTES:

Chapter 4: WIREFRAME AND SURFACE DESIGN FUNDAMENTALS

Introduction

Chapter 4 focuses on CATIA's *Wireframe and Surface Design* workbench. The reader will learn how to create basic wireframe surfaces and use them to create solids.

Tutorials Contained in Chapter 4

- Tutorial 4.1: Wireframe toolbar
- Tutorial 4.2: Surfaces toolbar
- Tutorial 4.3: Operations toolbar

NOTES:

Chapter 4:
WIREFRAME AND SURFACE DESIGN FUNDAMENTALS

Tutorial 4.1: Wireframe toolbar

<u>Featured Topics & Commands</u>

<u>Prerequisite Knowledge & Commands</u>

- Entering workbenches
- The *Sketcher* workbench and associated commands
- The *Part Design* workbench and associated commands

The Wireframe and Surface Design Workbench

The *Wireframe and Surface Design* workbench is used to create complex wire frame profiles and surfaces. In the *Part Design* workbench we worked under a *PartBody* which contained features used to create a solid. In the *Wireframe and Surface Design* workbench an *Open body* will contain features used to create surface elements. When creating reference elements (planes, lines, points) in *Part Design*, an *Open body* is automatically created to contain these elements. All the elements drawn in the *Wireframe and Surface Design* workbench will be under an *Open body*. Nothing will be solid. At any time you may insert an *Open body* by selecting <u>I</u>nsert – O<u>p</u>en Body.

The general process for creating a wireframe object is as follows: enter the *Wireframe and Surface Design* workbench, create wireframe geometry, create surfaces from existing geometry, modify and transform the surfaces (trim, join), enter the *Part Design* workbench, create a part body (solid), and modify the part using *Part Design* commands.

The *Wireframe and Surface Design* workbench contains the following standard workbench specific toolbars.

- <u>Wireframe toolbar:</u> This toolbar contains commands used to create reference elements such as points, lines, and curves that will be used to create surfaces.

- <u>Surfaces toolbar:</u> Once the reference geometry is constructed, the commands located in the *Surfaces* toolbar are used to create the surfaces.

- <u>Operations toolbar:</u> The commands located in the *Operations* toolbar are used to modify and transform surfaces and reference elements.

- <u>Replications toolbar:</u> The commands located in the *Replications* toolbar are used to duplicate features such as points and planes.

The Wireframe toolbar

The commands located in the *Wireframe* toolbar are used to create reference elements such as points, lines, and curves that will be used to

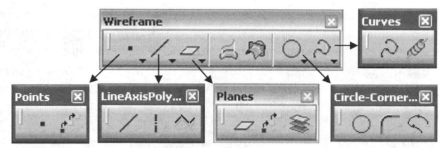

build surfaces. The commands and sub-toolbars located in the *Wireframe* toolbar, reading left to right, are

- The Points toolbar: Contains command to create reference points relative to the origin, another point, a curve or a surface.
- The LineAxisPolyline toolbar: Used to create lines, axes, and polylines.
- Plane: Allows you to create reference planes similar to that in the *Part Design* workbench. You can use planes as reference elements or cutting elements.
- Projection: This command creates a projection of a point, curve, or surface onto a support element, usually a plane or surface.
- Intersection: This command creates a curve that follows the intersection of a surface with a specified element.
- The Circle-Corner toolbar: Contains commands that will create circles, corners (fillets), and connect curves.
- The Curves toolbar: Used to create splines and helixes.

The Points toolbar

A point can be defined by its coordinates from a reference point (origin or selected point), or with respect to an element. You can edit any point by double clicking on it or its identifier in the specification tree. The commands located in the *Points* toolbar, reading left to right, are

- Point: The *Point* command allows you to create a point by entering coordinates relative to the origin or another point. It also allows you to create a point on a curve or surface relative to a point (end point, corner, reference point) on the curve or surface.

- Points Creation Repetition: The *Points Creation Repetition* command enables you to create several instances of a point.

The LineAxisPolyline toolbar

Lines can be used as guides, axes, or directions. They can also be joined to other elements. Lines can be created from points or vertices. They can be created on curves or surfaces. Lines have a direction associated with them depending on how you choose your start and end point. You can edit any line by double clicking on it or its identifier in the specification tree. The commands located in the *LineAxisPolyline* toolbar, reading left to right, are

- Line: The *Line* command allows you to create lines in various ways. A few examples are; between two points, from a point extending in a specified direction, and normal to a plane.
- Axis: The *Axis* command creates a reference element that is usually used to indicate symmetry.
- Polyline: The *Polyline* command allows you to create multiple lines that are connected together.

The Planes toolbar

Planes are reference elements that aid in the drawing process. Planes can be created from points or lines. They can be created on curves or surfaces. You can edit any plane by double clicking on it or its identifier in the specification tree. The commands located in the *Planes* toolbar, reading left to right, are

- Plane: The *Plane* command allows you to create a reference plane in a variety of ways.
- Points and Planes Repetition: The *Points and Planes Repetition* command creates multiple points and planes along a curve at a specified interval.
- Planes Between: The *Planes Between* command allows you to create any number of planes between two existing planes in one step.

The Circle-Corner toolbar

The *Circle-Corner* toolbar contains commands that are used to create curved elements. The commands, reading from left to right, are

- Circle: Used to create circles. Unlike drawing a circle in the *Sketcher* workbench, a support plane or surface must be specified. If the surface is not flat, the circle will be projected normally onto the surface.
- Corner: This command creates a fillet between two elements.
- Connect Curve: This command creates a curve that will connect two elements.

The Curves toolbar

Curves are used as guides, reference elements, or as limits to surfaces. The commands located in the *Curves* toolbar are used to create complex curves. The commands, reading from left to right, are

- Spline: A spline is a fitted curve that connects several points. Tangencies may be specified at its ends.
- Helix: A helix is a spiral that travels along a specified axis.

Tutorial 4.1 Start: Part Modeled

The part modeled in this tutorial is shown on the right. Reference elements (points, lines, planes, splines, ect...) are used to define surfaces. Once a surface model is created in the *Wireframe and Surface* workbench, the part is taken to the *Part Design* workbench and solids are created.

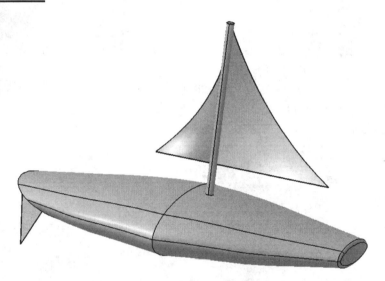

Section 1: Points and lines

1) Open a **New Part** drawing and name it *sail board.*

2) Enter the *Wireframe and Surface Design* workbench .

3) Save your drawing as *T4-1.CATPart*.

4) Set your length units to be millimeters (**Tools – Options... – Parameters and Measures**)

5) Insert a point at the origin. Select the **Point**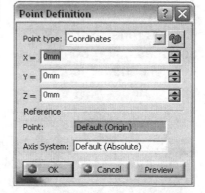
icon. In the *Point Definition* window fill in the
following fields:
- `Point type`: **Coordinates**
- X = *0 mm*, Y = *0 mm*, Z = *0 mm*
- `Reference Point`: **Default (Origin)**.

6) Create a line that starts at the origin and travels

   100 mm in the z direction. Select the **Line**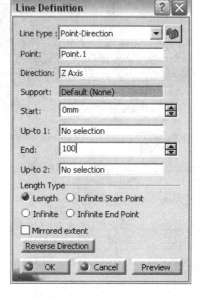
   icon. In the *Line Definition* window fill in the
   following fields:
- `Line type`: **Point-Direction**
- `Point`: **Point.1** (the point at the origin)
- `Direction`: **Z Component** (right click to
  access the directions)
- `Support`: **Default (None)**
- `Start`: *0 mm*
- `End`: *100 mm*
- `Length type`: **Length**.

7) Create a point that is (-10, 30, -50) millimeters
away from the top end of the line. Select the

   Point 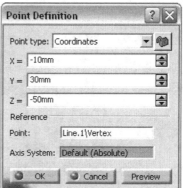 icon. In the *Point Definition* window fill
   in the following fields:
- `Point type`: Coordinates
- X = *-10 mm*, Y = *30 mm*, Z = *-50 mm*
- `Reference Point`: **Line.1\Vertex** (The
  point at the top end of the line. This is not the
  end at the origin.).

8) Create points at (*0, 60, -70*), (*-10, -15, -50*), and (*0, -30, -70*) millimeters away
from the same end of the line.

9) Create a line the starts at Point.3 (see figure) and is perpendicular to the Line.1. Select the **Line**

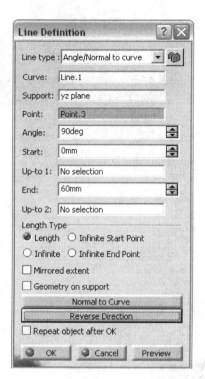

icon. In the *Line Definition* window fill in the following fields:

- Line type: **Angle/Normal to curve**
- Curve: **Line.1**
- Support: **yz plane**
- Point: **Point.3**
- Angle: ***90 deg***
- Start: ***0 mm***
- End: ***60 mm***
- Length type: **Length**
- Use the **Reverse Direction** icon if necessary.

10) Create a similar **Line** that starts at Point.5 (see figure). Use the **Reverse Direction** icon if necessary. (Line length is 30 mm)

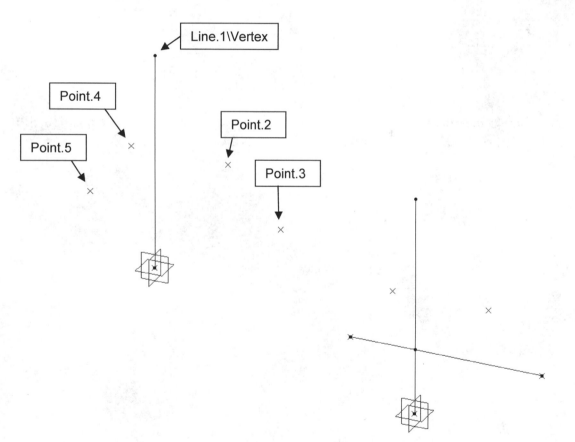

Section 2: Splines

1) Create a **Spline**  between **Line.1\Vertex**, **Point.2**, and **Point.3**. (Refer to the figure on the previous page.) Create a similar **Spline** between **Line.1\Vertex**, **Point.4**, and **Point.5**.

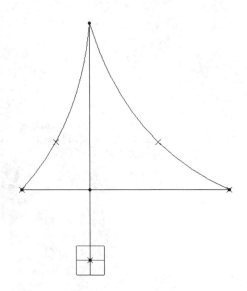

2) Create four **Points** that are located at (**50, -10, 0**), (**-50, -10, 0**), (**0, -10, 15**), and (**0, -10, -15**) millimeters relative to the origin.

3) Create a **Spline** using the 4 newly created points. In the *Spline Definition* window make sure to activate the **Close Spline** toggle, only after you have selected all the supporting points, to create a smooth closure.

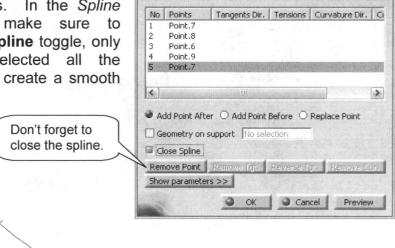

Don't forget to close the spline.

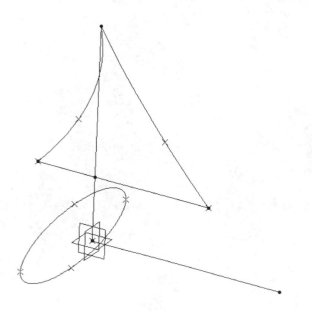

4) Draw a **Line** that starts at the origin and travels *100 mm* in the **y direction**.

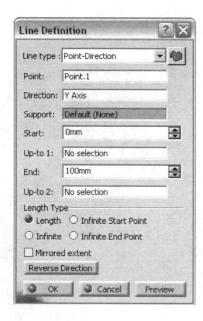

5) Create 3 equally spaced point/plane pairs along the newly created line. Select the **Points and Planes Repetition** 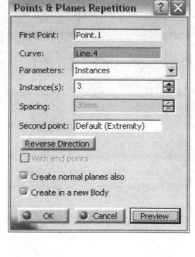 icon. In the *Points & Planes Repetition Definition* window fill in the following fields:

- `First Point`: **Point.1** (the point at the origin)
- `Curve`: **Line.4** (the newly created line)
- `Parameters`: **Instances**
- `Second Point`: **Default (Extremity)**
- Activate the **Create normal planes also** toggle.

Select **OK.**

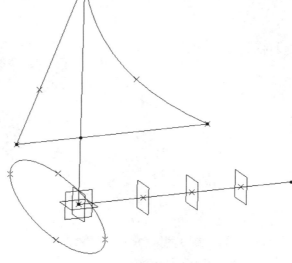

6) **Save** your drawing.

7) Project the ellipse located near the origin on to each of the newly created planes. Select the

Projection 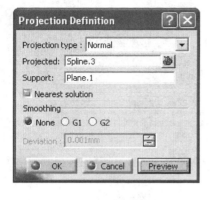 icon. In the *Projection Definition* window fill in the following fields:
- Projection type: **Normal**
- Projected: **Spline.3**
- Support: **Plane.1**

Follow a similar procedure for the remaining planes.

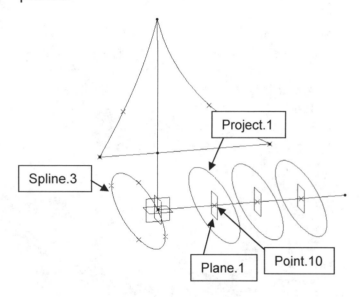

8) Scale the ellipse associated with Plane.1 by

80%. Select the **Scale** 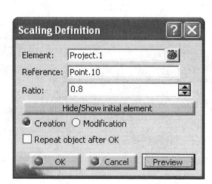 icon. This icon is located in the *Transformations* sub-toolbar which is located in the *Operations* toolbar. In the *Scaling Definition* window fill in the following fields:
- Element: **Projection.1**
- Reference: **Point.10**
- Ratio: *0.8*.

9) Using a similar process, scale the second projection by 60% and the third projection by 40%.

10) Insert points at the top of each of the three scaled

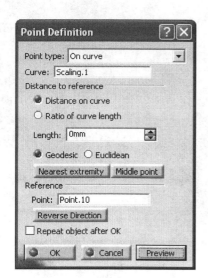

ellipses. Select the **Point** ■ icon. In the *Point Definition* window fill in the following fields:

- `Point type`: **On curve**
- `Curve`: **Scaling.1**
- `Distance to reference` **Distance on curve**
- `Length`: *0 mm*
- `Reference Point`: **Point.10** (The point at the center of the ellipse.)

Use a similar procedure to place a point on the top of the two remaining scaled ellipses.

(**PROBLEM?** Is the point at the bottom of the ellipse? Select the **Ratio of curve** length and enter *0.5 mm* in the ratio field.)

11) Create a **Spline** that connects the 5 points indicated in the figure.

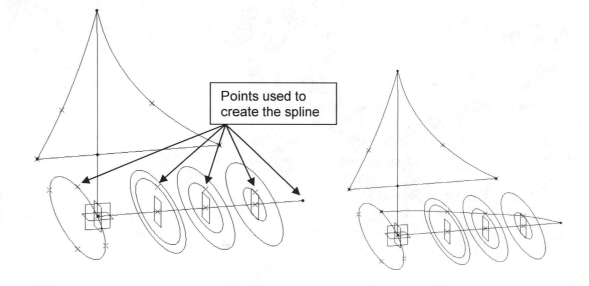

Points used to create the spline

12) Draw a *2mm* radius **Circle** ◯ centered at the origin and supported by the **xy plane**. To make a complete circle click on the **whole circle** icon.

13) **Save** your drawing.

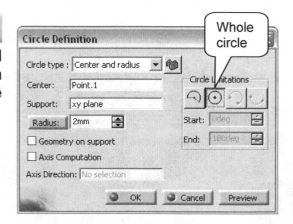

Section 3: Surfaces

1) Create surfaces for the sails. Select the **Fill**

 icon. This icon is located in the *Surfaces* toolbar. Select **Spline.1, Line.2,** and **Line.1** (see figure). Use a similar procedure to create a surface between **Spline.2**, **Line.3**, and **Line.1**.

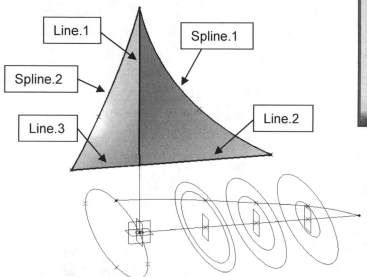

2) Activate the *Graphic Properties* toolbar. At the top pull down menu, select **View – Toolbars – Graphic Properties**. The *Graphic Properties* toolbar may be used to change the color of your surfaces. Try to change the color of your sails.

3) Create a mast by sweeping the circle drawn at the origin along the vertical line. Select the

   Sweep icon. This icon is located in the Surfaces toolbar. In the *Swept Surface Definition* window fill in the following fields:
   - `Profile:` **Circle.1**
   - `Guide curve:` **Line.1**.

4) Use the **Fill** command to fill the top and bottom of the mast with a surface.

5) Use the ellipses and spline to create half of the sail board. Select the **Multisections**

   Surface icon. Select the 4 ellipses starting with the original ellipse and selecting the other scaled ellipses in order of creation. In the *Multi-sections Surface Definition* window, select the **Spine** tab. Activate the `Spine:` field and select the spline that connects the 4 ellipses.

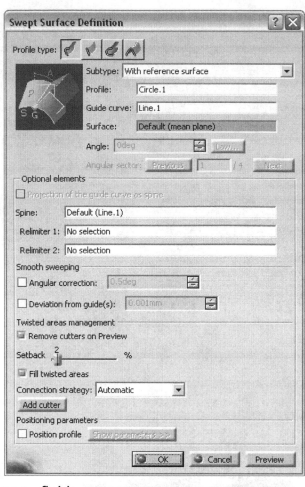

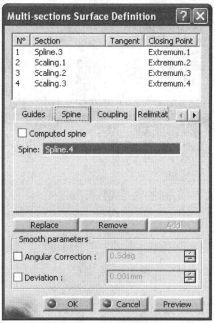

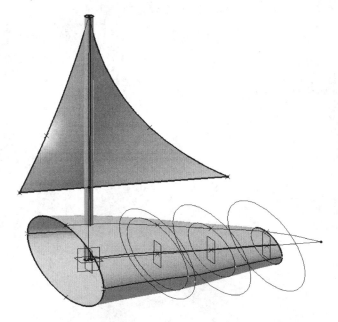

6) Use the **Fill** 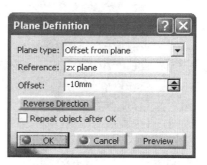 command to fill the small open end of the surfboard with a surface.

7) Use the **Plane** command to offset the **zx plane -10 mm** in the y direction. The plane should be in the same location as the big open end of the surfboard.

8) Mirror the sail board using the newly created plane as the mirroring surface.

Select the **Symmetry** icon. This icon is located in the *Transformations* toolbar. In the *Symmetry Definition* window activate the Element: field and select the sail board (**Multi-sections Surface.1**). Activate the Reference: field and select the newly created plane (**Plane.4**).

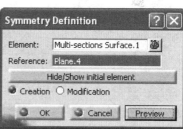

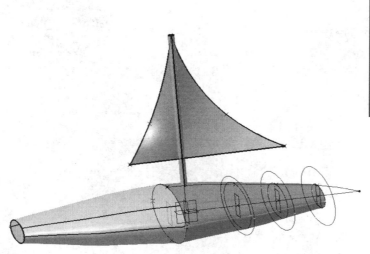

9) Create a point on the top of the mirrored small end

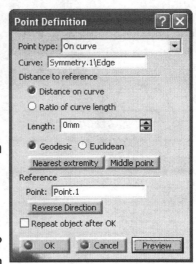

of the sail board. Select the **Point** icon. In the *Point Definition* window fill in the following fields:
- Point type: **On curve**
- Curve: **Symmetry.1\Edge**
- Distance to reference **Distance on curve**
- Length: *0 mm*
- Reference Point: **Point.1**

(**PROBLEM?** Is the point at the bottom of the ellipse? Select the **Ratio of curve** length and enter *0.5 mm* in the ratio field.)

10) Create two **Points** 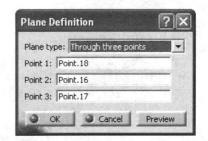 located at (*10, 20,-10*), and (*-10, 20, -10*) millimeters relative to the point at the top-end of the sail board (the last point created).

11) Use the last three points created to define a reference plane. Select the **Plane** icon. In the *Plane Definition* window select the Plane type: to be **Through three points** and then select the three points.

12) Use the newly created plane to create an intersection curve with the mirrored end of the sail board. Select the **Intersection** icon. In the *Intersection Definition* window select the mirrored end of the surfboard (**Symmetry.1**) as the First Element:, and the newly created plane (**Plane.5**) as the Section Element:.

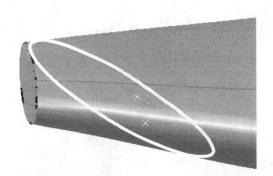

13)**Save** your drawing.

14)**Fill**  the intersection curve with a surface.

(**PROBLEM?** If the intersection curve does not fill, the most common reason is that the original spline used to create the initial ellipse was not closed.   This usually produces a gap at the top of the intersection curve.   Access the original ellipse and close it.   If you get an error, undo the action and use a *Line* to connect the gap and try to *Fill* it again.)

15)Use the intersection surface to trim off the end of the surfboard. Select the **Trim** icon.   This   icon   is   located   in   the *Operations* toolbar.   In the Trim Definition window select the mirrored end of the surfboard (**Symmetry.1**) as Element 1:, and   the   intersection   surface   (**Fill.6**)   as Element 2:.   (If the *Trim* command is trimming off the wrong side, select the *Other side* button.)

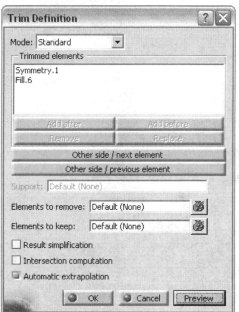

16)If the intersection curve lost its surface, **Fill** it again.

17)Create a point on the surface of the angled end of the sail board (the intersection surface).   Select the **Point** icon.   In the *Point Definition* window fill in the following fields:
- Point type: **On plane**
- Plane: select the angled plane used to create the intersection (Plane.5).
- H: *0 mm*
- V: *70 mm*
- Reference Point: **Default (Origin)**.
- Projection Surface: **Default (None)**.

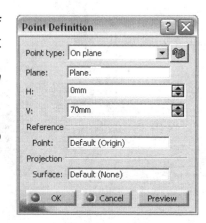

18) Create a line that starts at the newly created point and travels out from the surface 30 mm in a normal direction. Select the **Line** 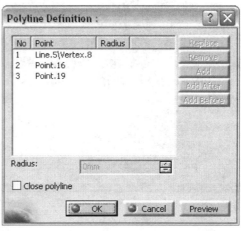 icon. In the *Line Definition* window fill in the following fields:

- Line type: **Normal to surface**
- Surface: select the angled end of the surfboard
- Point: select the newly created point
- Start: *0 mm*
- End: *30 mm*
- Length Type **Length**.
- You may have to use Reverse Direction to get the line to travel away from the surfboard.

19) Create a **Polyline** using the ends of the newly created line and the point indicated in the figure.

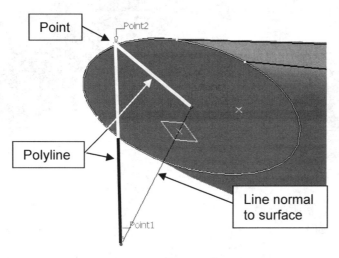

20) **Fill** the triangular curve with a surface.

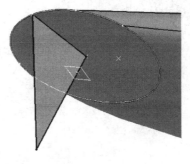

21)**Join** 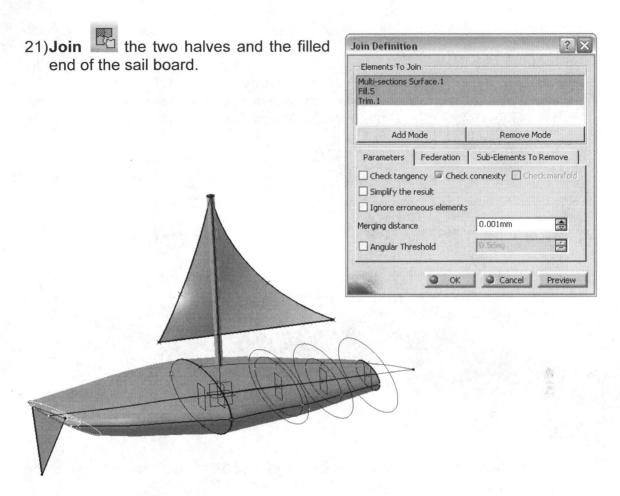 the two halves and the filled end of the sail board.

22)**Save** your drawing.

Section 4: Creating solids

1) Enter the **Part Design** workbench.

2) Activate the **Surface-Based Features (Extended)** toolbar if it is not already active.

3) Create a solid using the joined surface used to define the sail board. Select

the **Close Surface** 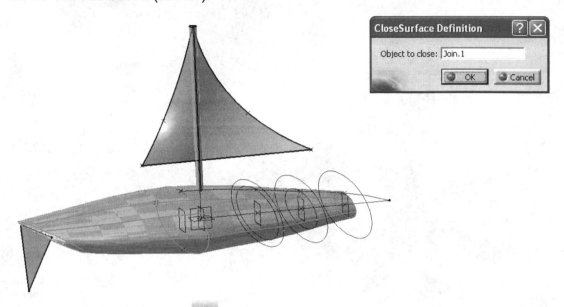 icon. Select the sail board and click **OK** in the *CloseSurface Definition* window. Notice that the surface of the sail board may now look checkered. This is because both the solid and the surface now exist in the same space. Go to the specification tree and hide the surface used to define the sail board (Join.1).

CloseSurface Definition [?][X]

Object to close: Join.1

[OK] [Cancel]

4) Use the **Close Surface** command to create a solid out of the swept mast. Then, hide the surface (Sweep.1).

5) We no longer need to see the reference elements (points, lines, planes, etc..). You may hide all of these if you like.

6) Now that you have solid, you may apply any *Part Design* workbench specific command to them. Apply a *1.5 mm*

radius **Edge Fillet** to the end of the surfboard that does not have the fin.

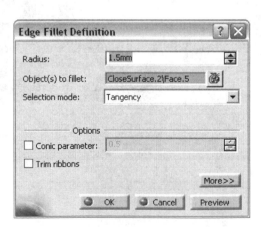

Edge Fillet Definition [?][X]

Radius: 1.5mm

Object(s) to fillet: CloseSurface.2\Face.5

Selection mode: Tangency

Options

☐ Conic parameter: 0.5

☐ Trim ribbons

[More>>]

[OK] [Cancel] [Preview]

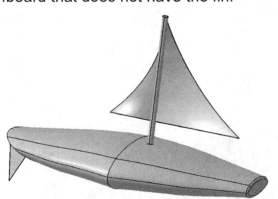

Chapter 4:
WIREFRAME AND SURFACE DESIGN FUNDAMENTALS

Tutorial 4.2: Surfaces toolbar

Featured Topics & Commands

Prerequisite Knowledge & Commands

- Entering workbenches
- The *Sketcher* workbench and associated commands
- The *Part Design* workbench and associated commands
- The *Wireframe* toolbar

The Surface toolbar

The commands located in the *Surfaces* toolbar are used to create surfaces that are defined by one or more reference elements. The commands, reading from left to right, are

- Extrude: An extruded surface is created from an open or closed curve. Its direction and limits are specified.

- Revolve: A revolved surface is created from an open or closed curve. The axis and angle of rotation are specified.

- Sphere: The *Sphere* command allows you to create a complete or incomplete sphere. Limits for the sphere may be specified in both the revolved direction (meridian) and the top to bottom direction (parallel).

- Cylinder: The *Cylinder* command allows you to create a complete or incomplete cylinder.

- Offset: An offset surface is created from an existing surface. An offset distance and direction are specified. The resulting surface is parallel to the existing one.

- Sweep: A swept surface is created from an open or closed curve that is swept along guide curves.

- Fill: The filled surface is created between joined curves. The shape of the resulting surface depends on the continuity between the support surfaces (if any) and the fill surface.

- Multisections Surface: A multisections surface or lofted surface is defined by several parameters including; tangency, closing point, coupling, and guide curve.

- Blend: The blend surface is created between two curves. The curves used must be single edge curves and not closed. The shape of the resulting surface depends on the continuity between the support surfaces (if any) and the blend surface.

Tutorial 4.2 Start: Part Modeled

The part modeled in this tutorial is created with reference elements and surfaces. The surfaces are created in a variety of ways in order to gain experience with as many commands in the *Surfaces* toolbar as possible.

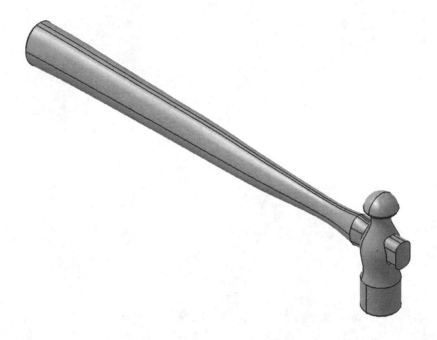

Section 1: Multisections surface (Loft)

1) Open a **New... Part** drawing, name it *Hammer* and enter the *Wireframe and Surface Design* workbench .

2) **Save** your drawing *T4-2.CATPart*.

3) Set your length units to be inches (**Tools – Options... – Parameters and Measures**)

4) Create 5 **Points** that are located at (*0.5, 0, 0*), (*-0.5, 0, 0*), (*0, 0, 0.625*), (*0, 0, -0.625*), and (*0, 11, 0*) relative to the **origin**.

5) Create 4 **Points** 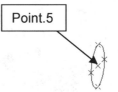 that are located at (*0.3125, 0, 0*), (*-0.3125, 0, 0*), (*0, 0, 0.4375*), (*0, 0, -0.4375*) relative to the **Point.5** (the point located 11 inches from the origin).

6) Use a **Spline** to connect the first 4 points and last 4 points as shown in the figure. Start and end the spline at the top and activate the **Close spline** toggle.

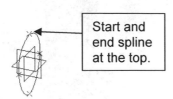

Start and end spline at the top.

Point.5

7) Create 12 **Points** that are located at (*0, 2, 0.594*), (*0, 4, 0.531*), (*0, 6, 0.5*), (*0, 8, 0.406*), (*0, 9, 0.375*), (*0, 10, 0.344*), (*0.4375, 2, 0*), (*0.406, 4, 0*), (*0.375, 6, 0*), (*0.3125, 8, 0*), (*0.25, 9, 0*), (*0.219, 10, 0*) relative to the **origin**.

8) Use a **Spline** to connect the first 6 points and last 6 points with the points on the two ellipses as shown in the figure.

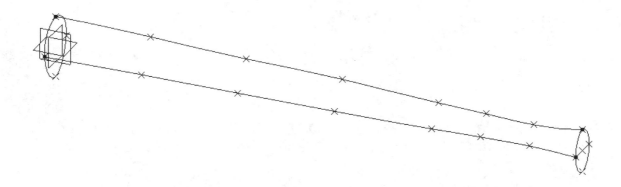

9) Use a multisections surface to loft together the two ellipses using the two long splines as guides. Select the **Multisections Surface** 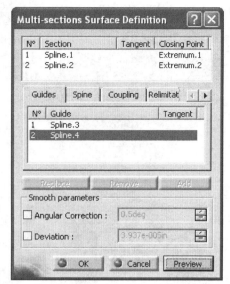 icon. In the *Multi-sections Surface* Definition window, activate the *Section* area and select the two ellipse (Spline.1 and Spline.2). Activate the *Guide* area and select the two long splines (Spline.3 and Spline.4). Make sure both closing points are at the top.

10) **Save** your drawing.

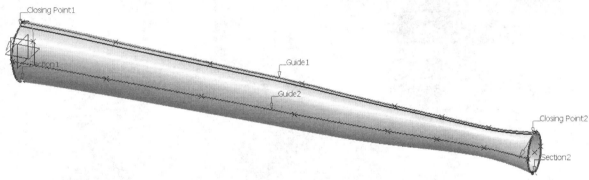

Section 2: Extruded surface

1) Create 4 **Points** ▪ that are located at (*0.25, 0.5, 0.375*), (*-0.25, 0.5, 0.375*), (*0.25, 0.5, -0.375*), (*-0.25, 0.5, -0.375*) relative to the **Point.5** (the point located 11 inches from the origin).

2) Draw 4 **Lines** ╱ connecting the last 4 points drawn as shown in the figure.

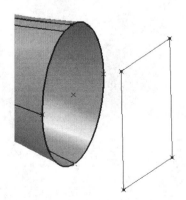

3) Create rounded corners or fillets at each vertex of the rectangle. Select the **Corner** icon located under the *Circle* icon. Select two adjoining lines to be Element 1: and Element 2:. Activate both **Trim element** toggles and enter a Radius: of *0.19 in*. Repeat this procedure for the next two corners. On the last corner you need to activate the **Corner On Vertex** toggle.

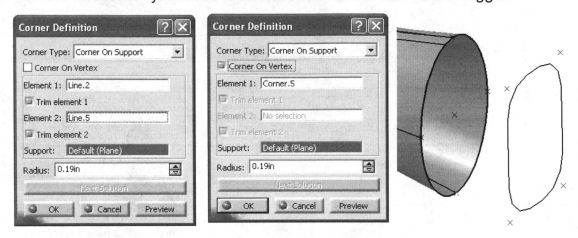

4) **Extrude** the cornered rectangle by *1 in* along the **Y Axis**.

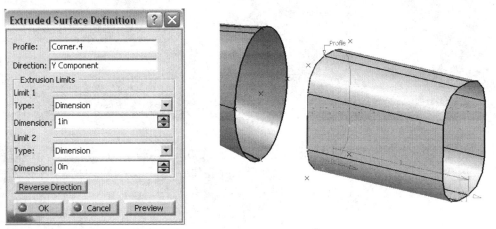

5) **Save** your drawing.

Section 3: Blended surface

1) Create a surface that connects Multi-sections surface.1 and Extrude.1. Select the **Blend** icon. In the *Blend Definition* window, activate the First curve: field and select the spline used to create the end of Multi-sections surface.1 (**Spline.2**). Activate the Second curve: field and select the cornered rectangle. Note where the *First Point* and *Second Point* lie. They should be across from each other to prevent twist. If your *Second Point* is not across from your *First Point,* click on the **Closing Points** tab and **Replace** your *Second Point* with a point that is across from the *First Point.*

Need to move the *Second Point* to be across from the *First Point.*

The *Second Point* is now across from the *First Point.*

2) Notice that the interface between Blend.1 and Extrude.1 is discontinuous or sharp. Make that interface smooth. Get back into the *Blend Definition* window by double clicking on **Blend.1** in your specification tree. Activate the *Second support* field and select **Extrude.1**. (A gap WARNING window may appear. Just click OK.)

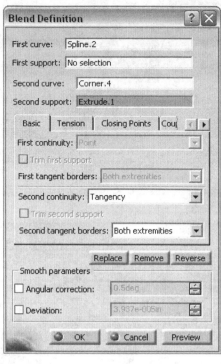

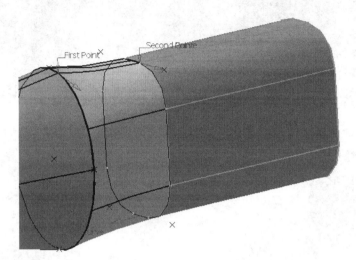

3) **Fill** 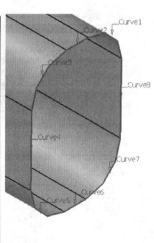 the two ends of the handle.

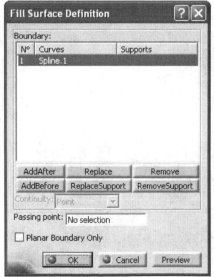

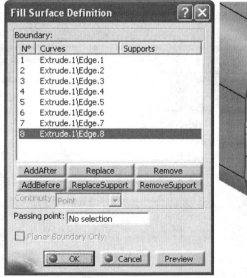

4) **Join** all surfaces of the handle.

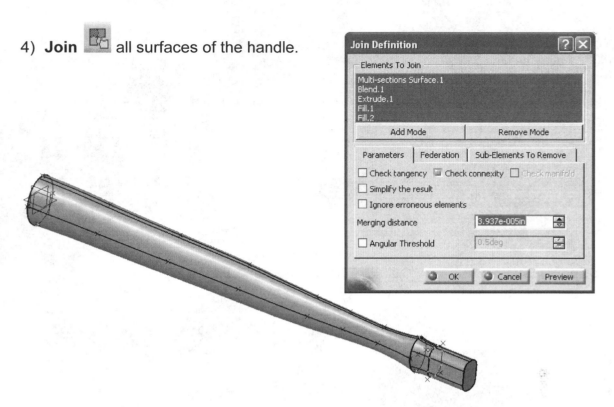

Join Definition

Elements To Join

Multi-sections Surface.1
Blend.1
Extrude.1
Fill.1
Fill.2

Add Mode	Remove Mode

Parameters | Federation | Sub-Elements To Remove

☐ Check tangency   ☑ Check connexity   ☐ Check manifold
☐ Simplify the result
☐ Ignore erroneous elements

Merging distance   3.937e-005in

☐ Angular Threshold   0.5deg

● OK   ● Cancel   Preview

Section 4: Spheres and Cylinders

1) Insert a new Open Body (**Insert – Body**).

2) Insert a new **Plane** that is offset from the **zx plane** by **11.75 in**.

Plane Definition

Plane type: Offset from plane
Reference: zx plane
Offset: 11.75in
Reverse Direction
☐ Repeat object after OK

● OK   ● Cancel   Preview

3) Insert a **Point** at (**0, 11.75, 0**) inches relative to the origin. Then insert a **Point** at (**0, 0, -1**) inches relative to the previous point.

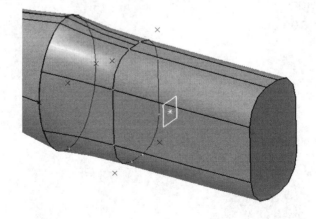

4) Select the **Cylinder** 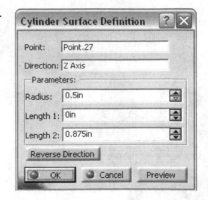 icon. In the *Cylinder Surface Definition* window fill in the following fields;

- Point: **Point.27** (most recently created point)
- Direction: **Z Component**
- Radius: *0.5 in*
- Length 1: *0 in*
- Length 2: *0.875 in*.

5) Insert a **Point** at (*0, 0, 1*) inches relative to the point in the middle of the offset plane (Point.26).

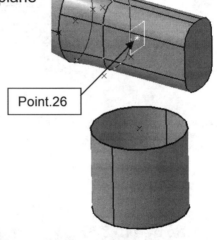

Point.26

6) Select the **Sphere** icon. In the *Sphere Surface Definition* window, fill in the following fields;

- Center: **Point.28** (most recently created point)
- Sphere radius: *0.438 in*
- Select the incomplete sphere icon
- Parallel Start Angle: *0 deg*
- Parallel End Angle: *90 deg*
- Meridian Start Angle: *0 deg*
- Meridian End Angle: *360 deg*.

7) **Fill** 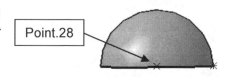 the bottom of the cylinder. The end that is furthest away from the handle.

8) **Save** your drawing.

Section 5: Revolved surface

1) Insert 2 **Points** 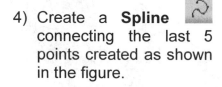 at (*-0.5, 0, 0*) and (*-0.44, 0, 0.25*) inches relative to the center point of the cylinder (Point.27).

2) Insert 2 **Points** at (*-0.438, 0, 0*) and (*-0.25, 0, -0.25*) inches relative to the center point of the sphere (Point.28).

3) Insert a **Point** at (*-0.5, 0, 0*) inches relative to the point located in the middle of the offset plane (Point.26).

4) Create a **Spline** connecting the last 5 points created as shown in the figure.

5) Create a **Line** that connects the center of the cylinder with the center of the sphere.

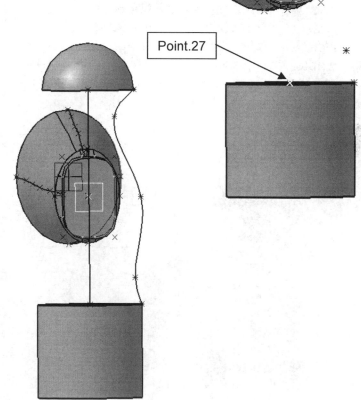

6) **Revolve** 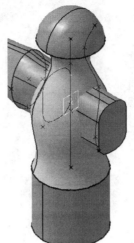 the spline *360 deg* about the line.

7) Create an **Intersection** 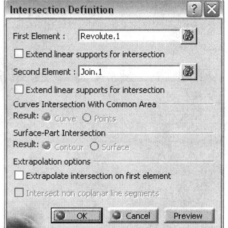 curve between the revolved surface and the handle. When you select the **OK** button a *Multi-Result Management* window will appear. This means that two intersections were created and we need to tell CATIA which one to keep. After you select the **OK** button, a *Near Definition* window will appear. CATIA wants you to choose an element that is near the intersection curve that you want to keep. Select an edge of the handle end.

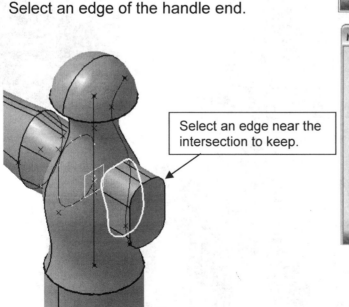

Select an edge near the intersection to keep.

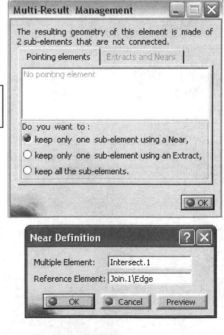

8) Hide the handle. Right click on **PartBody** and select **Hide/Show**.

9) **Extrude** 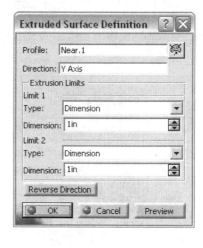 the intersection curve (**Near.1**) *1 inch* in both directions along the **Y Axis/Component**.

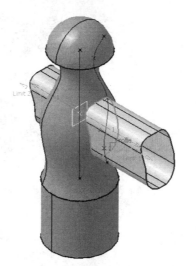

Extruded Surface Definition

Profile: Near.1
Direction: Y Axis

Extrusion Limits
Limit 1
Type: Dimension
Dimension: 1in
Limit 2
Type: Dimension
Dimension: 1in

Reverse Direction

OK   Cancel   Preview

10) **Trim** 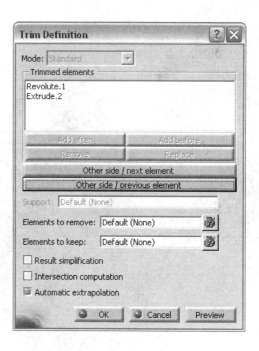 **Revolute.1** and **Extrude.2** to create a hole going through the hammer head. Select the **Other side / previous elements** buttons until the desired result is obtained.

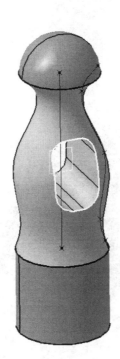

Trim Definition

Mode: Standard

Trimmed elements
Revolute.1
Extrude.2

Add after   Add before
Remove   Replace
Other side / next element
Other side / previous element

Support: Default (None)

Elements to remove: Default (None)
Elements to keep: Default (None)

☐ Result simplification
☐ Intersection computation
☑ Automatic extrapolation

OK   Cancel   Preview

11)**Join**  all the surfaces of the hammer head.

12)Show the hammer handle and hide all reference elements.

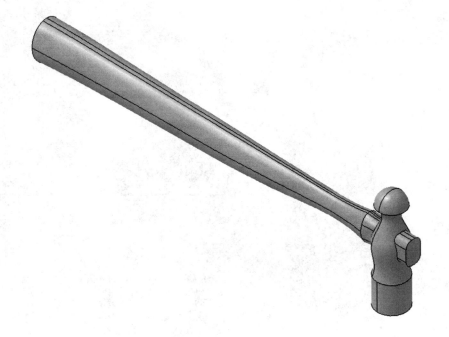

13)**Save** you part.

Section 6: Creating solids

1) Enter the **Part Design** 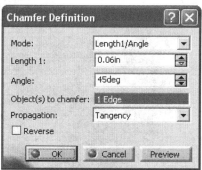 workbench.

2) Right click on **PartBody** and select **Define in Work Object**.

3) Use the **Close Surface** command to create a solid out of the handle (Join.1).

4) Hide the surfaces of the handle (Join.1).

5) Apply a **0.06 in x 45 deg Chamfer** to the end of the hammer handle. The end away from the hammer head.

Chamfer Definition

Mode:	Length1/Angle
Length 1:	0.06in
Angle:	45deg
Object(s) to chamfer:	1 Edge
Propagation:	Tangency
☐ Reverse	

OK   Cancel   Preview

6) Right click on **Body.2** and select **Define in Work Object**.

7) Use the **Close Surface** command to create solids out of the hammer head (Join.2).

8) Hide the surfaces of the hammer head (Join.2).

9) Apply a **0.06 in** radius **Fillet**  to the striking surface of the hammer head and the edge of the ball, as shown in the figure.

10) Make the handle **Wood** and the hammer head **Steel**.

Chapter 4:
WIREFRAME AND SURFACE DESIGN FUNDAMENTALS

Tutorial 4.3: Operations toolbar

Featured Topics & Commands

Prerequisite Knowledge & Commands

- Entering workbenches
- The *Sketcher* workbench and associated commands
- The *Part Design* workbench and associated commands
- The *Wireframe* and *Surfaces* toolbar

The Operations toolbar

The *Operations* toolbar contains commands the will help you transform and modify surfaces and curves and make it easier to produce the desired shape. The commands and sub-toolbars located in the *Operations* toolbar, reading left to right, are

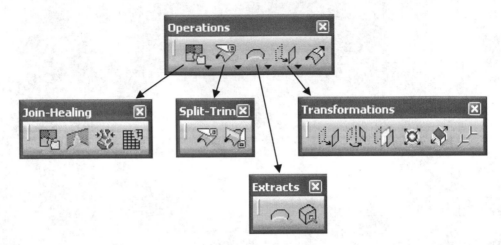

- <u>Join-Healing toolbar:</u> Contains commands that will join, fill gaps, restore, and disassemble adjoining elements.
- <u>Split-Trim toolbar:</u> Contains commands that will cut one side of intersecting elements.
- <u>Extracts toolbar:</u> Contains commands that create curves and surfaces from surface boundaries and edges.
- <u>Transformations toolbar:</u> Contains commands that will transform the size, position, orientation, etc... of an element.
- <u>Extrapolate:</u> The *Extrapolate* command extends one element to the boundary of another.

The Join-Healing toolbar

The commands contained in the *Join-Healing* toolbar, reading left to right, are

- <u>Join:</u> The *Join* command allows you to join one or more adjacent elements into a single element. Multiple curves or surfaces may be joined. You can define a *Merge distance* and an *Angle tolerance*. The *Merge distance* is the maximum gap distance under which the two elements can be joined. The *Angle tolerance* specifies the maximum interface angle under which the two elements can be joined.
- <u>Healing:</u> The *Healing* command fills in gaps between two elements. A *Merge distance* or maximum gap that will be filled, and a *Distance objective* or minimum gap that will be filled, may be specified.
- <u>Untrim Surface or Curve:</u> The *Untrim Surface or Curve* command allows you to rebuild an element that has been split one or more times.

- Disassemble: The *Disassemble* command breaks a multi-celled surface or curve into its individual cells.

The Split-Trim toolbar

The commands contained in the *Split-Trim* toolbar, reading left to right, are

- Split: The *Split* command cuts or splits one element using a cutting element. One side of the element is cut.
- Trim: The *Trim* command cuts or trims two intersecting elements. One side of each element is cut.

The Extracts toolbar

The commands contained in the *Extracts* toolbar, reading left to right, are

- Boundary: The *Boundary* command creates a curve from a surface boundary. Limits may or may not be set.
- Extract: The *Extract* command creates a curve from a surface edge. It can also be used to extract a surface. Limits may not be set.

The Transformations toolbar

The commands contained in the *Transformations* toolbar, reading left to right, are

- Translate: The *Translate* command makes a duplicate element that is a specified distance away from the original. Multiple duplicates may be created using the repeat option.
- Rotate: The *Rotate* command makes a duplicate element that is rotated about an axis by a specified angle. Multiple duplicates may be created using the repeat option.
- Symmetry: The *Symmetry* command mirrors an element about a reference element.
- Scaling: The *Scaling* command makes a duplicate element that is of a different size from the original. Multiple duplicates may be created using the repeat option.
- Affinity: The *Affinity* command creates a scaled version of an element. However, the scale factors in each direction may be different.
- Axis to Axis: The *Axis to Axis* command takes an element that is supported by one axis and creates a duplicate that is supported by another axis.

The Grid toolbar

The *Grid* toolbar is located in the *Tools* toolbar. The commands located in the *Grid* toolbar, reading left to right, are

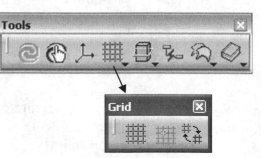

- <u>Work on Support:</u> This command allows you to work on a sketch plane.
- <u>Snap to Point:</u> This command allows you to snap to the grid points.
- <u>Working Supports Activity:</u> This command exits the working support mode.

Tutorial 4.3 Start: Part Modeled

The part modeled in this tutorial is a combination of shapes used to illustrate the commands found in the *Operations* toolbar. The reference elements used to constructed the base where drawn using the *Work on Support* command. Several commands found in the *Operations* toolbar and its sub-toolbars are used to modify and add to the base.

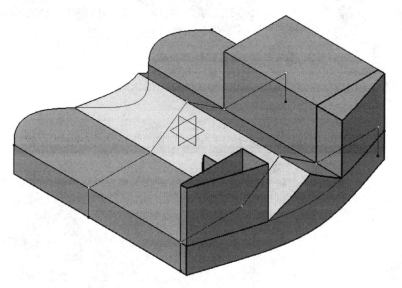

Section 1: Using Supports

1) Open a **New... Part** drawing, name it **Sheet Metal** and enter the *Wireframe and Surface Design* workbench .

2) **Save** your drawing as **T4-3.CATPart**.

3) Set your length units to be millimeters (**Tools – Options... – Parameters and Measures**)

4) Pull out the **Grid** toolbar . This sub-toolbar is located in the *Tools* toolbar which is usually in the bottom toolbar area.

5) Select the **Work on Support** icon. In the first *Work on Support* window select the **zx plane** as the Support:. In the second *Work on Support* window set the Primary spacing: to be *100 mm* and the Graduations: to be **10**.

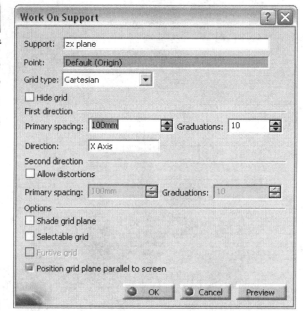

6) Activate the **Snap to point** icon.

7) Draw the following profile using the **Polyline** command.

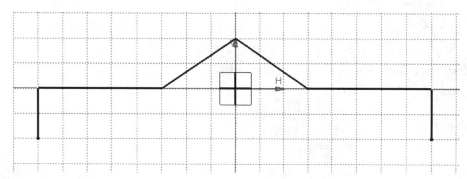

8) Select the **Works Support Activity** icon to deactivate the support. Then, select the **Isometric View** icon.

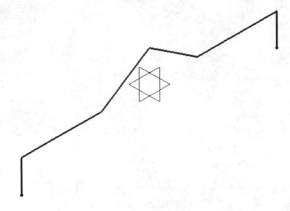

9) **Extrude** the polyline **75 mm** in the positive and negative **y** directions. (Direction: Y Component)

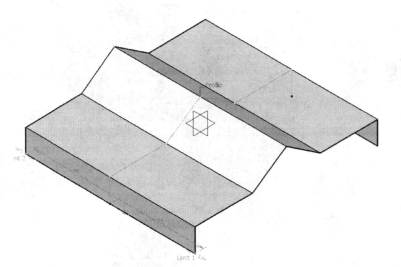

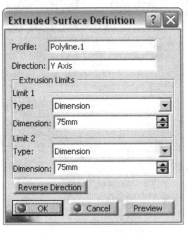

10) Use the **Work on Support** command to sketch on the **xy plane** with the same grid spacing as before.

11) Hide Extrude.1.

12) Draw the following 2 **Splines** 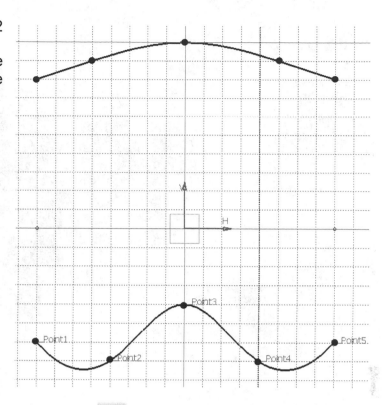 using the points indicated in the figure.

13) Unhide Extrude.1.

14) Select the **Works Support Activity** icon to deactivate the support. Then, select the **Isometric View** icon.

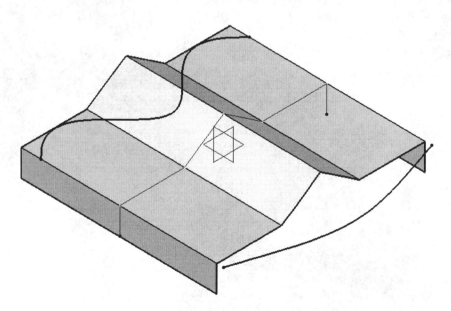

15) **Extrude** 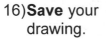 each spline **20 mm** in the both the negative and positive **z** directions.

16) **Save** your drawing.

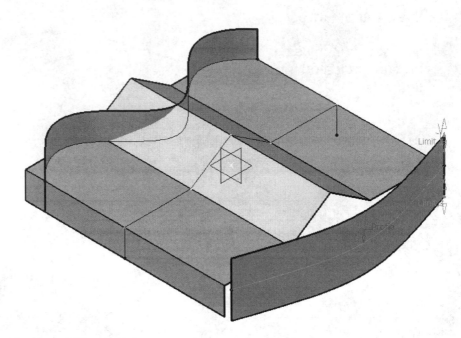

Section 2: Split, Trim, and Extrapolate

1) Select the **Split** 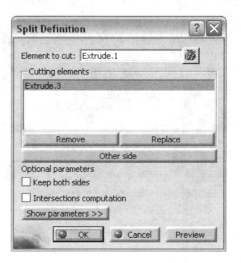 icon. In the *Split Definition* window, activate the Element to cut: field and select **Extrude.1**. Activate the Cutting elements area and select the extrude created with the wavy spline. Use the **Other side** button if necessary to get the desired results. Notice that the cutting element does not change.

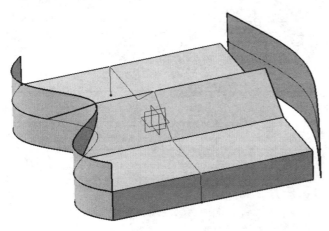

2) Delete Split.1 and select the **Trim** 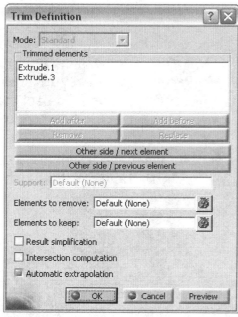 icon. Select Extrude.1 and Extrude.3 (the wave spline extrude). Use the **Other side/ previous elements** buttons if necessary. Notice that both elements get cut.

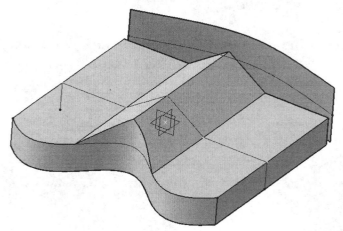

3) Create a boundary to use in the *Extrapolate* command. (Note: The figure shows Extrude.2 hidden) Select the **Boundary** icon. In the *Boundary Definition* window, fill in the following fields;

- `Propagation type`: **Point continuity**
- `Surface edge`: select the original extrude now called **Trim.1**
- `Limit 1`: select **Point 1** as shown on the figure
- `Limit 2`: select **Point 2**
- We want the boundary to cover the original polyline shape, if it does not, click on the arrow attached to Point 1.

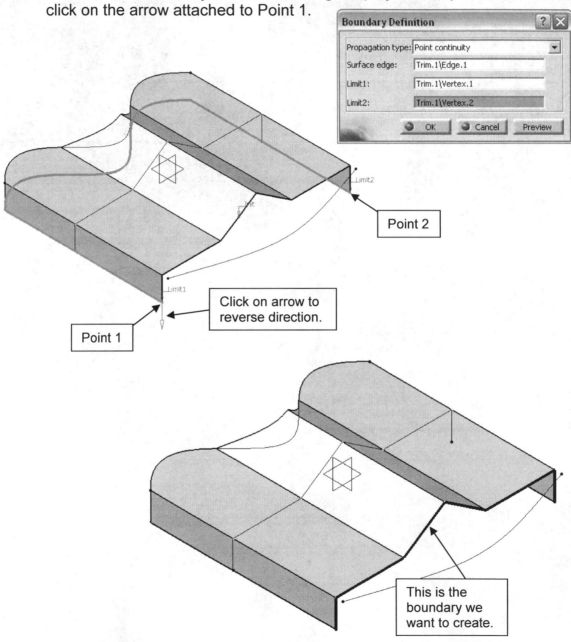

4) Select the **Extrapolate** 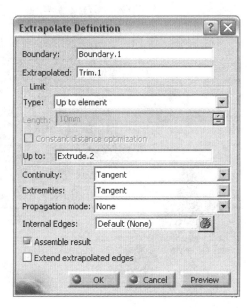 icon. In the *Extrapolate Definition* window, fill in the following fields;
   - Boundary: **Boundary.1**
   - Extrapolate: **Trim.1**
   - Type: **Up to element**
   - Up to: select the extrude created from the curved spline.

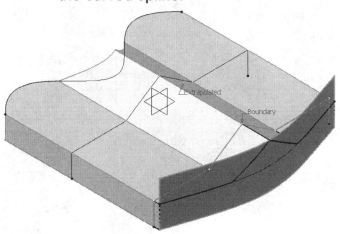

5) Use the **Split** command to achieve the results shown in the figure.

6) **Save** your drawing.

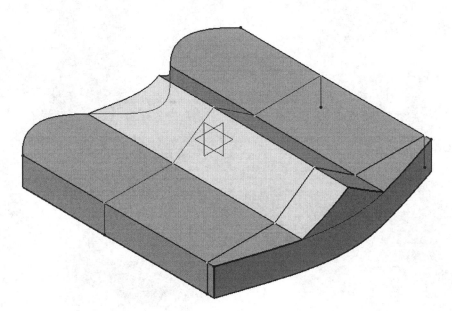

Section 3: Extracting

1) Select the **Extract** ⬚ icon and select the face indicated in the figure.

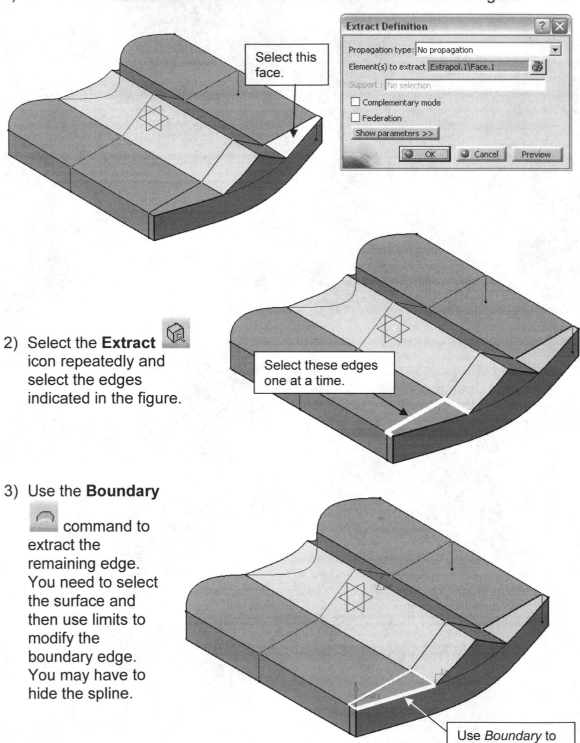

Select this face.

Extract Definition

Propagation type: No propagation

Element(s) to extract: Extrapol.1\Face.1

Support : No selection

☐ Complementary mode

☐ Federation

Show parameters >>

OK   Cancel   Preview

2) Select the **Extract** icon repeatedly and select the edges indicated in the figure.

Select these edges one at a time.

3) Use the **Boundary** ⌒ command to extract the remaining edge. You need to select the surface and then use limits to modify the boundary edge. You may have to hide the spline.

Use *Boundary* to extract this edge.

4) **Join**  all the edges created in the previous 2 steps.

5) **Extrude** 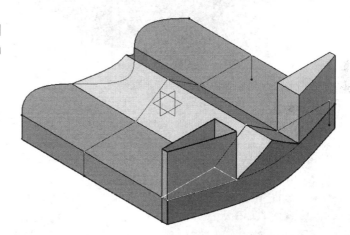 **Extract.1** and **Join.1** individually *40 mm* in the **z** direction.

6) Disassemble the closed extrude so that one of its surfaces can be extruded. Select the closed extrude (Extrude.4 or Extrude.5 depending on the order in which you extruded) then select the

   Disassemble  icon. A *Disassemble* window will appear showing how many elements the extrude will be broken up into. Select **OK**.

7) **Extrude**  the back surface of the disassemble extrude **75 mm** in the **y** direction as shown in the figure. (**Problem?** If you get an error, select the surface in the specification tree.)

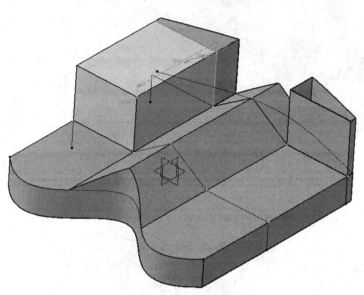

8) Use the Affinity command to create a scaled duplicate of the open extrude. Select the **Affinity** icon. In the *Affinity Definition* window, fill in the following fields;

- `Element:` select the open extrude (Extrude.4 or Extrude.5)
- `Origin:` select the point indicated in the figure
- X: *0.5*, Y: *-1.5*, Z: *0.5*.

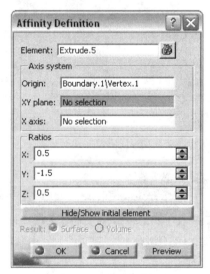

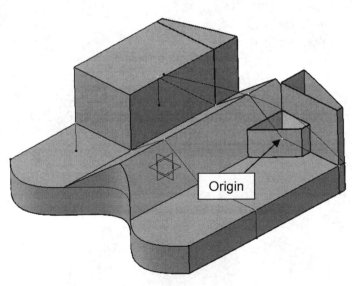

Chapter 4: Exercises

Exercise 4.1: This exercise can be performed after completing the tutorials presented in chapter 4. Model the following parts.

   a) A pair of pliers
   b) A screwdriver
   c) A popup tent
   d) An airplane body
   e) A car body

NOTES:

Chapter 5: ASSEMBLY DESIGN FUNDAMENTALS

Introduction

Chapter 5 focuses on CATIA's *Assembly Design* workbench. The reader will learn how to create and manage an assembly or product, and apply assembly constraints.

Tutorials Contained in Chapter 5

- Tutorial 5.1: Assembly Constraints and Advanced Commands
- Tutorial 5.2: Advanced Assembly

NOTES:

NOTES:

Chapter 5: ASSEMBLY DESIGN FUNDAMENTALS

Tutorial 5.1: Assembly Constraints and Advanced Commands

<u>Featured Topics & Commands</u>

<u>Prerequisite Knowledge & Commands</u>

- Entering workbenches
- The *Sketcher* workbench and associated commands
- The *Part Design* workbench and associated commands
- Moving and rotating parts with the compass

Assembly Design Workbench

The *Assembly Design* workbench allows you to design in the context of an assembly. You can add new or existing parts, or sub-assemblies to the root assembly. Parts can be positioned and constrained within the assembly. You can also analyze interferences between parts.

Product Structure

The product structure is best visualized by looking at the specification tree. Let's take, for example, the assembly of a machine vise. At the root is the product or assembly (Machine Vise). Under the assembly are parts (Base, Sliding Jaw, Jaw Plate, etc...), a subassembly (Handle Assembly), and constraints that are used to place parts in their functional positions. Notice that there are multiple copies of a single part. For example, there are 4 instances of Part11. Part11 was not drawn four different times; it was merely copied and pasted using the familiar CTRL+C and CTRL+V shortcuts. The root assembly and parts can be named by right clicking on the name and accessing the *Properties* window.

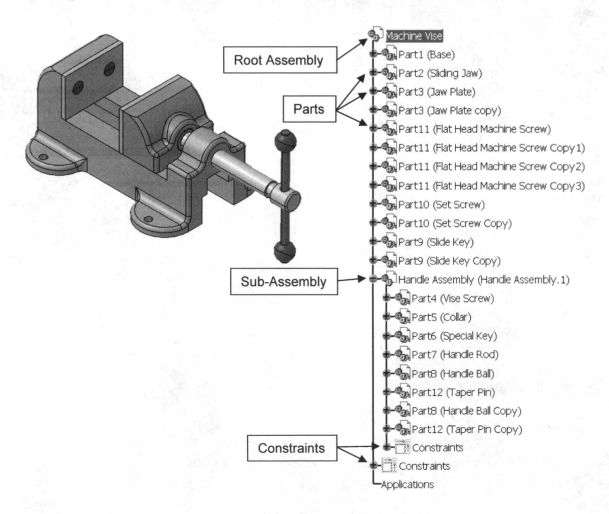

The Product Structure Tools toolbar

This toolbar allows you to insert, replace, and reorder components in your product. When you are inserting an existing part or subassembly, their corresponding files are not copied into the assembly. They are just referenced by the assembly. The commands located in the *Product Structure Tools* toolbar, reading left to right, are

- Component: This command inserts a new component that exists only in the root assembly file (CATProduct) and not as its own file.
- Product: This command allows you to insert a new product or subassembly (CATProduct) into the root assembly.
- Part: This command allows you to insert a new part (CATPart) that will be designed on the fly while you are designing the assembly.
- Existing Component: This command is used when you want to add an existing part or assembly as a component.
- Existing Component With Positioning: This command is used when you want to add an existing part or assembly as a component with a position.
- Replace Component: This command is used when you want to replace a component.
- Graph Tree Reordering: This command allows you to reorder your components in the specification tree.
- Generate Numbering: This command allows you to associate numbers or letters to each part of the assembly. These numbers or letters will be used when creating and ballooning the assembly drawing in the *Drafting* workbench.
- Selective Load: This command tells you which components in the assembly will be loaded (opened or active) or not, and components that will be hidden or not.
- Manage Representations: This command allows you to manage the representations (name, association, etc...) of the components in the assembly.
- Multi Instantiation toolbar: This toolbar contains commands that allow you to create instances or duplicates of a component.

The Catalogue toolbar

The *Catalogue Browser* command lets you browse through existing already drawn catalogue of standard parts such as bolts, screws and nuts. Therefore, there is no need to model these components if they can be retrieved from the catalogue.

The Constraints toolbar

Assembly constraints are the constraints used to place each part into the assembly in its functional position. The general process for constraining the parts in an assembly is; fix one component in space, use the compass to move the other parts into their approximate position, position the parts precisely using the correct constraint(s), and then update  the assembly to move the parts into position. If a part(s) has been moved out of position, the update command may be used to move the part(s) back into position. Constraint symbols may be hidden by right clicking on the *Constraints* inference in the specification tree and selecting *Hide/Show*. The commands located in the *Constraints* toolbar, reading left to right, are

- Coincidence Constraint: The *Coincidence Constraint* creates alignments between axes, planes, or points.
- Contact Constraint: The *Contact Constraint* creates contact between two planes or faces.
- Offset Constraint: The *Offset Constraint* defines an offset distance between two elements.
- Angle Constraint: The *Angle Constraint* defines an angle between two elements.
- Fix Component: This command fixes a part in space such that it is always located in the same spot. The other parts are constrained relative to the fixed part.
- Fix Together: This command fixes two parts together.
- Quick Constraint: This command automatically creates what CATIA believes is the correct constraint between two elements.
- Flexible/Rigid Sub-Assembly: This command allows you to change a sub-assembly between flexible and rigid. A flexible sub-assembly has parts or components that can be moved disregarding the fact that it is not the active component. The positions of the parts can be different than those stored in the reference CATProduct file.
- Change Constraint: This command allows you to change the type of constraint that exists between two elements.
- Reuse Pattern: This command allows you to reuse an existing pattern.

Save Management

Since assemblies contain several file links, the *Save Management...* command should be used to save it. A *Save Management* window will appear showing all the CATProducts and CATParts that have been or need to be saved. You can click on each individually and save the file using the *Save* or *Save As* commands. Also, the *Save All* command will save all documents that have been modified since the last save.

The Measure toolbar

The *Measure* toolbar contains commands that measure distances, angles, mass, volume, inertia, etc... of your assembly or individual geometric element. You can keep the result of your measurement as an element in your specification tree. The commands located in the *Measure* toolbar, reading left to right, are

- Measure Between: This command measures between two geometric elements.
- Measure Item: This command measures several physical parameters of an element such as area and perimeter.
- Measure Inertia: This command measures physical parameters of a part or assembly. It will calculate the mass, volume, moments of inertia, and center of gravity among other things.

The Space Analysis toolbar

The commands located in the *Space Analysis* toolbar, reading left to right, are

- Clash: This command analyzes any part interference. If there is an actual material interference between two parts, you will be warned.
- Sectioning: This command allows you to see a 2D section, and to create a 3D section of your assembly.
- Distance and Band Analysis: This command analyzes the minimum distance between two sections or between one component and all the other components.

The Annotation toolbar

Annotations are text and symbols that can be seen in the 3D view. The commands located in the *Annotations* toolbar, reading left to right, are

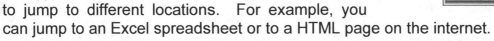
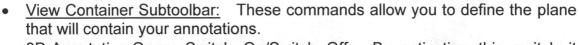

- <u>Weld Feature:</u> This command allows you to put welding symbols on your part.
- <u>Text With Leader:</u> This command produces a text annotation that points to a selected component or geometric feature.
- <u>Flag Note With Leader:</u> A flag note allows you to add hyperlinks to your document that can be used to jump to different locations. For example, you can jump to an Excel spreadsheet or to a HTML page on the internet.
- <u>View Container Subtoolbar:</u> These commands allow you to define the plane that will contain your annotations.
- <u>3D-Annotation-Query Switch On/Switch Off:</u> By activating this switch it enables you to see the relationships between the annotations and the geometries in which they refer to.

Assembly/Parts Modeled

The assembly modeled in this tutorial is used to cover assembly constraints and more advanced commands located in the *Assembly Design* workbench. Among other things we will apply material, calculate physical properties, analyze interferences, and create a section view.

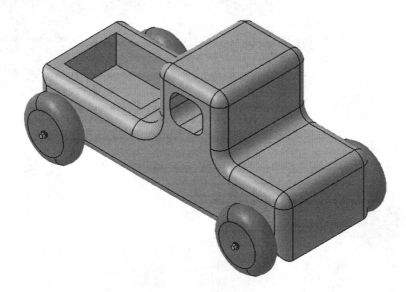

Section 1: Creating the assembly

1) Enter the **Assembly Design** 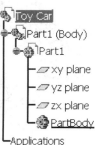 workbench. Notice that at the top of your specification tree there is a **Product**.

2) Set your units to **millimeters** (**Tools** – **Options...**).

3) Right click on **Product1** and select **Properties**. In the *Properties* window, click on the **Product** tab. In the `Part Number` field, rename your product ***Toy Car***.

4) Save your draw as ***Toy Car.CATProduct*** in a new folder called ***Toy Car***.

5) Select the **Part** 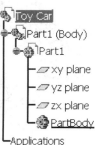 icon (If **Part1** does not appear in the specification tree, then click on **Toy Car**). This will add a new part to your assembly or product. Notice that *Toy Car* in the specification tree is highlighted in blue. This means that it is the active component.

6) Right click on **Part1 (Part1.1)** in the specification tree and select **Properties**. In the *Properties* window, click on the **Product** tab. In the `Part Number` field, type ***Part1***. In the `Instance Name` field, type ***Body***.

7) At the top pull down menu, select **File** – **Save Management...**. A *Save Management* window will appear indicating that *Toy Car.CATProduct* has been saved before, but modified since, and that *Body.CATPart* is new. Click on **Toy Car.CATProduct** and select the **Save** button. Then, click on **Part1.CATPart** and select the **Save As...** button. Name your file ***Part1– Body.CATPart*** and save it in the *Toy Car* folder. Select **OK**.

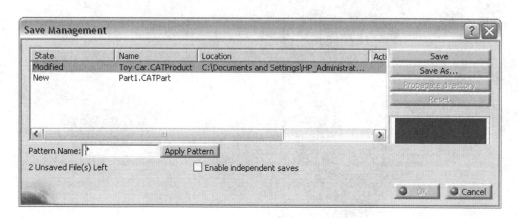

8) Expand *Part1 (Body)* in the specification tree until you can see its *PartBody*. Double click on **PartBody**. This will automatically enter you into the *Part Design* workbench. You are now ready to draw Part1. Notice that now Part1 is highlighted in blue. If CATIA does not enter the *Part Design* workbench, you will have to enter it manually (it may enter the *Wireframe* workbench if this was the last one you were in).

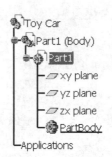

9) Draw the following **Profile** on the **yz plane** and **Pad** it to a length of *98 mm*.

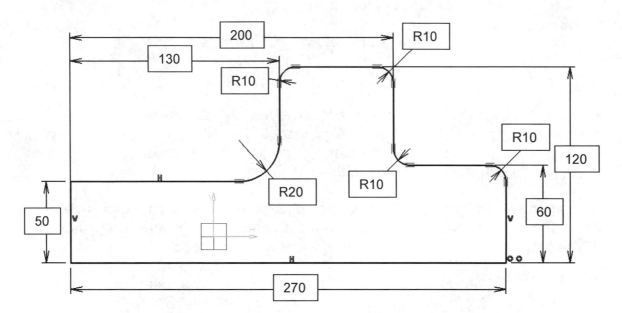

10) **Edge Fillet** the two side faces of the truck with a radius of *10 mm*.

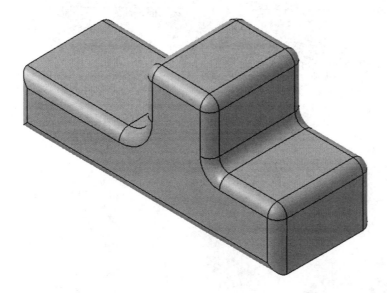

11) Draw and **Pocket** the following sketches.

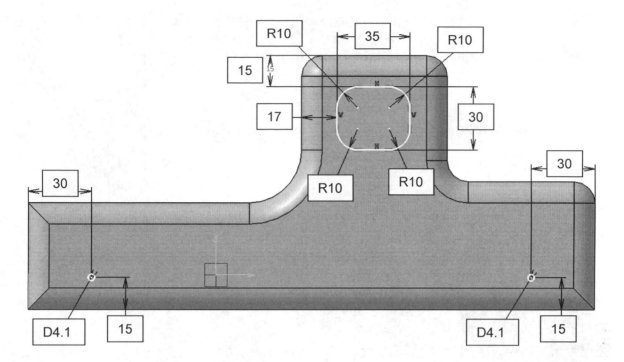

12) Enter the Sketcher on the top of the trucks bed. Project the face of the bed as a construction element and **Offset** it to the inside by *10 mm*. Turn the offset into a standard element. **Pocket** the offset to a depth of *30 mm*.

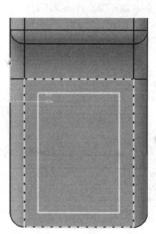

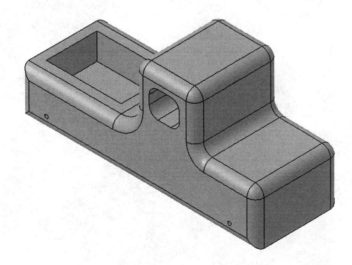

13) Make the *Body* plastic. Select **PartBody** in the specification tree and select the **Apply Material** 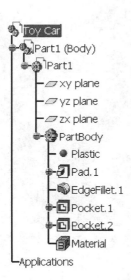 icon. In the *Library* window, select the **Other** tab and then **Plastic**. Select the **Apply Material** button, then **OK**.

14) Change the color of the body to red.

15) Use **Save Management** to save your product and part.

Section 2: Wheel Sub-Assembly

1) Double click on **Toy Car** in your specification tree to get back into the *Assembly Design* workbench.

2) Add a new **Product** 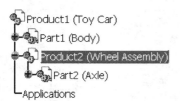 to your root assembly and name it **Wheel Sub-Assembly**. (Note: I renamed Toy Car to include *Product1* just for clarity.)

3) Add a new **Part** to your *Product2 (Wheel Assembly)* and name it **Axle**. When the *New Part: Origin Point* window appears, select **No**.

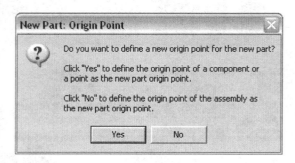

4) Double click on the **PartBody** under *Part2 (Axle)* to enter the *Part Design* workbench.

5) **Hide** *Part1 (Body)*.

6) Enter the **Sketcher** on the **zx plane** and sketch the axle. We will be creating a shaft that has symmetrical ends and is 150 mm long. Shown in the figure is one end.

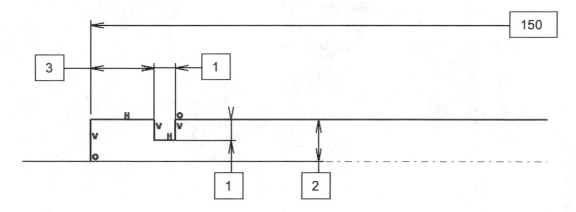

7) **Shaft** the *Axle* and make it **Steel**.

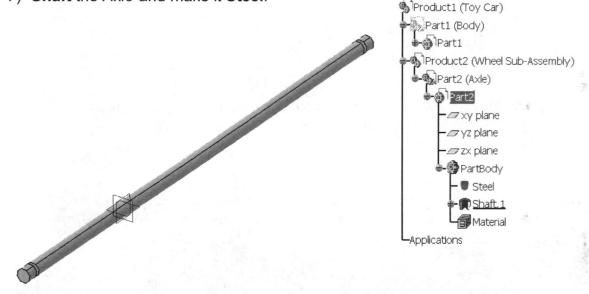

8) Double click on **Product2 (Wheel Sub-Assembly)**. Add a new **Part** to *Product2 (Wheel Sub-Assembly)* and name it **Wheel**.

9) **Save Management**. Save the *Wheel Sub-Assembly* as **Wheel Sub-Assembly.CATProduct**, the *Axle* as **Part2-Axle.CATPart** and the *Wheel* as **Part3-Wheel.CATPart**.

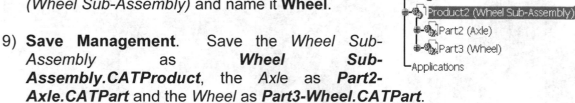

10) **Hide** *Part2 (Axle)*.

11) Double click on the **PartBody** under *Part3 (Wheel)* and sketch the following profile on the **zx plane**. **Shaft** the sketch, make the *Wheel* **Rubber**, and change its color to black.

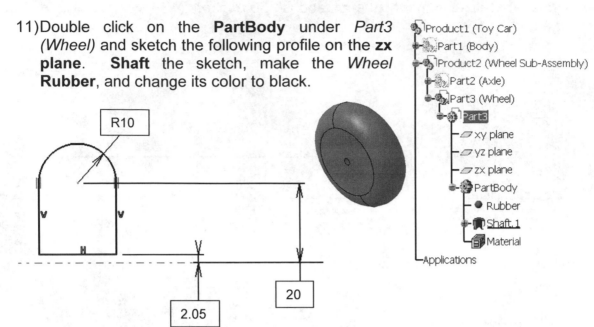

12) Double click on **Product2 (Wheel Sub-Assembly)**. Add a new **Part** to *Product2 (Wheel Sub-Assembly)* and name it **Snap Ring**.

13) **Hide** *Part3 (Wheel)*.

14) Double click on the **PartBody** under *Part4 (Snap Ring)* and sketch the following profile on the **yz plane**. **Pad** the sketch to a length *0.8 mm*, and make the *Snap Ring* **Steel.**

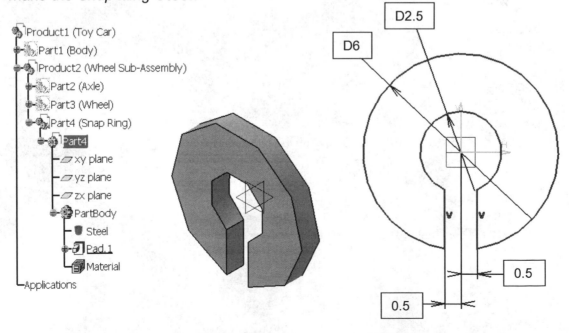

15)**Save Management**. Save the *Snap Ring* as ***Part4-Snap Ring.CATPart***.

16)**Show** all the parts under *Product2 (Wheel Sub-Assembly)*.

17)Double click on **Product2 (Wheel Sub-Assembly)** and then move the parts, using the compass, to approximately the positions shown.

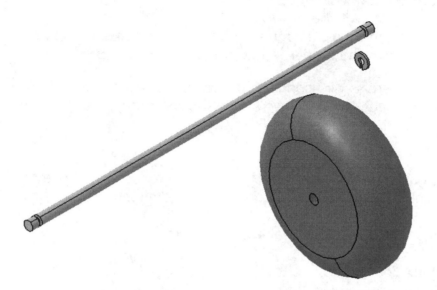

18)Make a copy of the *Wheel* and *Snap Ring* (using **copy** and **paste**) and move them to the positions shown. Name them as shown in the specification tree.

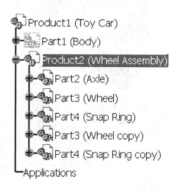

Section 3: Applying Constraints

1) Apply a coincidence constraints between the wheels, snap rings and the axle.

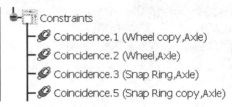

Select the **Coincidence Constraint** icon. If an *Assistant* window comes up, **Cancel** it. Select the axis of the *Axle* (you may have to move the mouse around a bit or even zoom in before it shows up) and then the axis the *Wheel*. A line with the coincidence constraint symbol at both ends will connect the two axes. Apply similar Coincidence Constraints between the *Axle* and the *Wheel Copy*, the *Axle* and the *Snap Ring* and the *Axle* and the *Snap Ring Copy*.

2) The part is not in position yet. Select the **Update All** icon to move the part into position. This icon is usually located in the bottom toolbar area.

3) Move the parts using the compass so that each is visible.

4) Apply a contact constraint between both snap rings and the axle. Select the

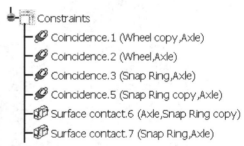

Contact Constraint icon. Select face of the *Snap Ring* and then the inside face of the *Axle* groove. A line with the contact constraint symbol at both ends will connect the two surfaces. Apply a similar constraint with the *Snap Ring Copy* on the other side of the *Axle*.

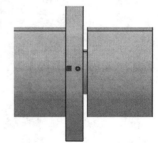

5) **Update All** .

6) Create an **Offset Constraint** between the *Wheels* and the *Snap Rings* of **1 mm**. A *Constraint Properties* window will appear. Fill in a value of *1. 0* mm in the Offset field.

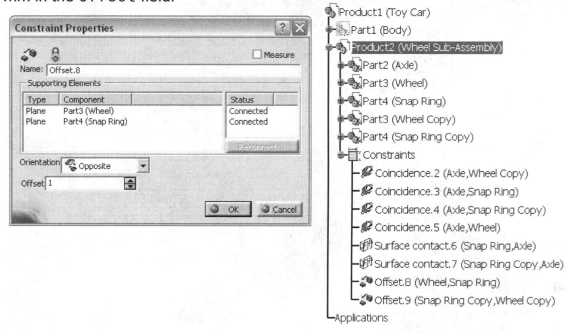

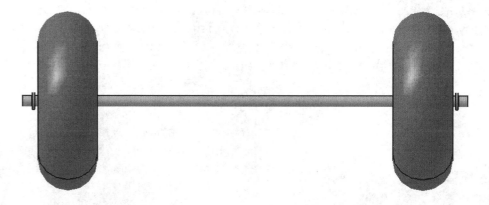

7) **Update All** .

8) Make a copy of the *Wheel Sub-Assembly*. Right click on **Product2 (Wheel Sub-Assembly)** and select **Copy**. Then, right click on **Toy Car** and select **Paste**. Move the sub-assembly with the compass as shown.   To move the *Wheel Sub-Assembly Copy* you need to be in the *Toy Car* product.

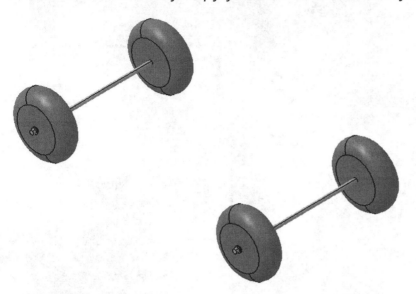

9) **Show** the *Body*.

10) Apply the appropriate constraints between the *Wheel Assemblies* and the *Body* as shown in the specification tree given. The **Coincidence** constraints are between the *Axle* and the holes in the *Body*. The **Offset** constraints are between the inside surface of the *Wheel* and the side of the *Body* and is 1 mm.

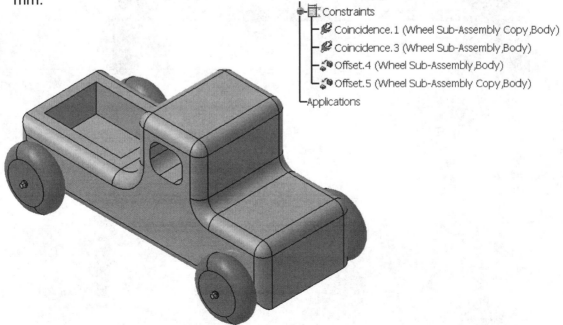

11) **Save Management.**

Section 4: Measuring

1) Double click on **Toy Car** to make it the active item.

2) Select the **Measure Between** icon. In the *Measure Between* window, select **Arc center** for both selection modes. Select each *Wheel*. A minimum distance of 210 mm will be calculated.

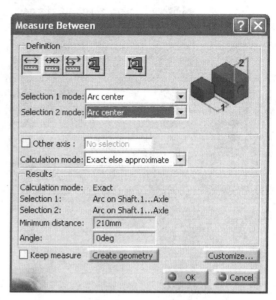

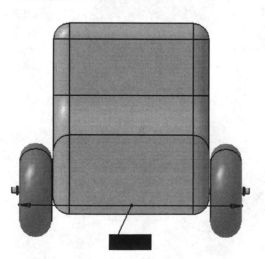

3) View the front of the car (use the *right side view* icon) and then select the

   Measure Between icon. In the *Measure Between* window, select **Edge only** for both selection modes. Select the outside surfaces of the *Wheels*. A minimum distance of 140 mm will be calculated.

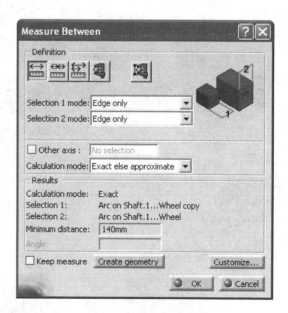

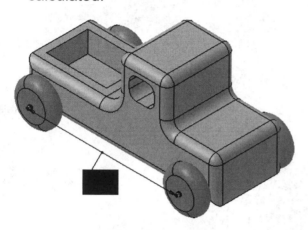

4) Select the **Measure Item** 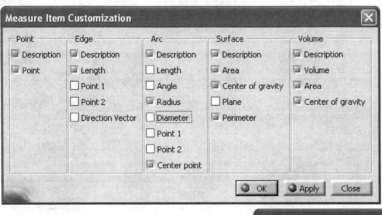 icon. In the *Measure Item* window, select the **Customize** button. In the *Measure Item Customization* window, activate the toggles shown and select **OK**. In the *Measure Item* window, select **Any geometry** as the selection mode and then select the front face of the car body. Notice that the surface area, center of gravity, and perimeter are calculated. (Note: The center of gravity may be different from what is shown. This is because we did not constrain the car body with respect to the global axis.)

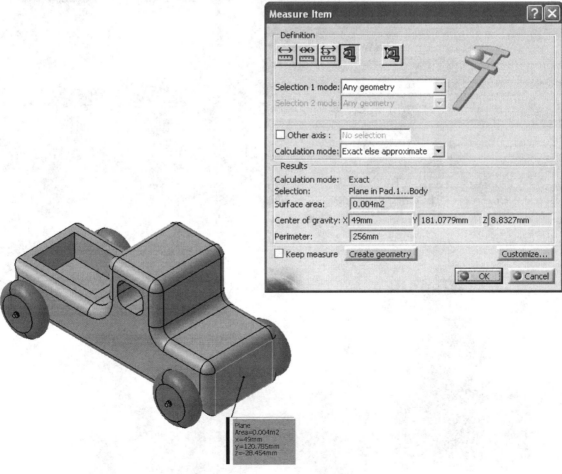

5) Select **Part1 (Body)** in the specification tree and then the **Measure Inertia** icon. Notice the physical parameters that are measured. In particular notice the center of Gravity. Select **Toy Car** in the specification tree and then the **Measure Inertia** icon. Does Gz increase or decrease? Does Gy increase or decrease? Why? Why does Gx not change?

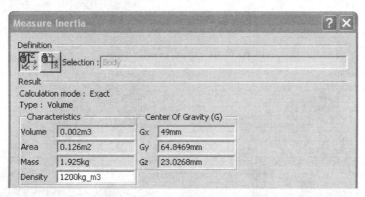

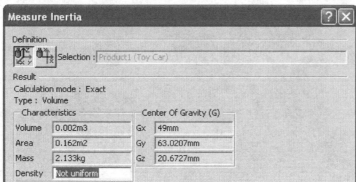

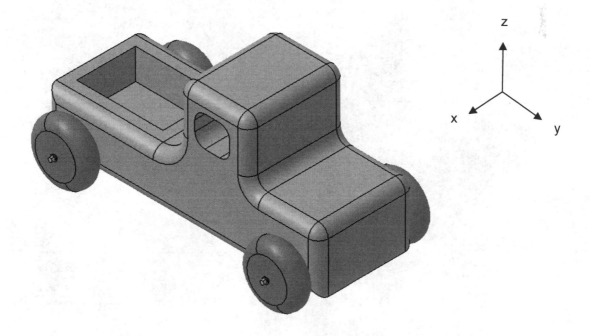

Section 5: Space Analysis

1) Make the root assembly active (*Toy Car*).

2) Select the **Clash**
icon. In the
Check   Clash
window set the type
to be **Contact +
Clash** and **Between

all   components**.
Select the **Apply**
button. Notice that
the window changes
and shows all the
components   that
contact one another
and shows no clashes.

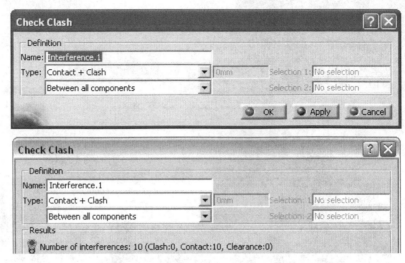

3) Edit the car *Body* and change one of the axle holes to a diameter of **3 mm**.

After you are done, re-enter the root assembly and re-apply the **Clash**
command. Notice that now there is one clash. Click on the clash and a
Preview window will appear showing you which two parts have an
interference.

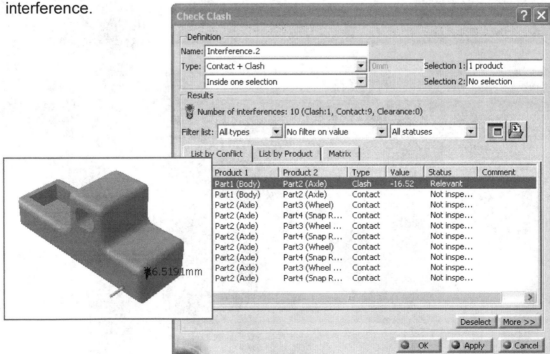

4) Change the axle hole diameter back to **4.1 mm**. After you are done, re-enter the root assembly.

5) Select the **Sectioning** 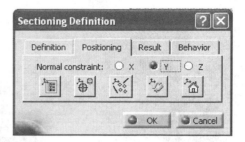 icon. A *Section.1* viewing window will appear and a sectioning plane will appear in the *Product* viewing window. In the *Sectioning Definition* window select the **Positioning** tab and activate the **Y** toggle. The size of the sectioning plane may be adjusted by selecting a border and moving it. The position of the sectioning plane may be changed by selecting inside the borders and moving it. Move the plane back and forth and notice that the section view changes. Change the positioning to **Z** and move the plane to see the section view change. Change the positioning to **X**. Select the coordinate axis in the middle of the sectioning plane and rotate it to see the effects (see figure on the next page). Re-select the **X** positioning and move the plane to the middle of the car. In the *Sectioning Definition* window, select the **Definition** tab. Activate the **Volume Cut** icon (see figure on the next page.) Select the **Cancel** button.

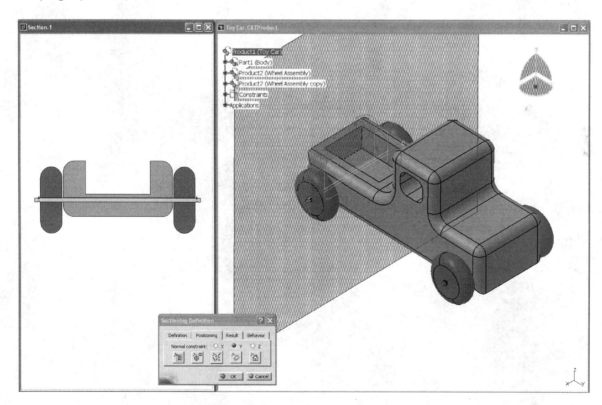

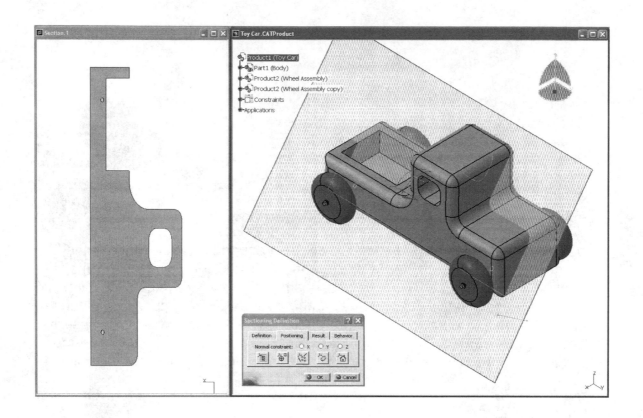

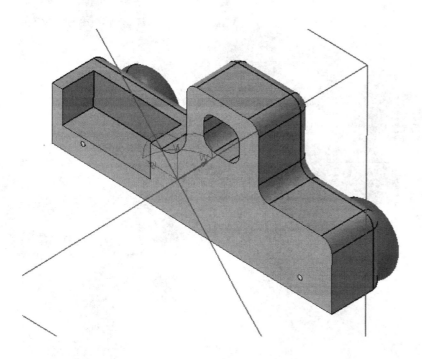

Section 6: Annotations and snap shots

1) Use the compass to create an exploded view of the toy car. First move the compass to one of the wheel assemblies and move it down. Repeat for the other wheel assembly. Try to move one *Wheel* off of the *Axle*. You will find that you can not. It is part of the *Wheel Assembly*. Double click on **Product2 (Wheel Assembly)**. Use the compass to move the *Wheel* and *Snap Ring* away from the *Axel* on one side of the car. Notice that both the *Wheels* and *Snap Rings* move together. Rotate the car so that you can manipulate the other side of the *Wheel Assembly*. Double click on **Toy Car**. Select the **Product2 (Wheel Assembly)** in the specification tree and then the

Flexible/Rigid  icon. Double click on **Product2 (Wheel Assembly)**. Now move the *Wheel* and *Snap Ring* that are in the *Wheel Assembly*. Now they move independently. Repeat for the *Wheel Assembly Copy*.

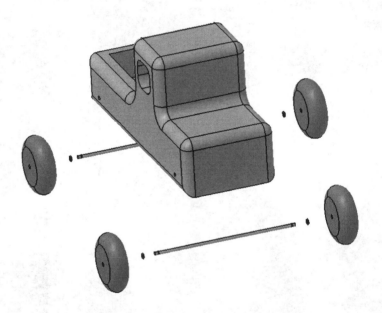

2) At the top pull down menu, select **Tools** – **Image** – **Capture**. In the Capture window, select the **Select Mode** icon. Draw a box around the exploded toy car and then select the **Capture** icon. A *Capture Preview* window will appear showing your snap shot. If this is what you want, select the **Save As** icon. In the *Save As* window, name your file and **Save**.

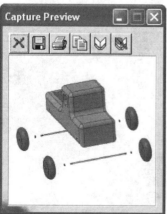

3) **Update All** (CTRL + U).

4) Select the **Text With Leader** icon and then select the front end of the car *Body*. In the *Text Editor* window, type **Body**. A text annotation will appear. This can be moved around and the dashed rectangle may be hidden. Right click on **Text.1 (Body)** in the specification tree and select **Properties**. In the *Properties* window select the **Font** tab. Set the font size to be *10 mm*. Create another annotation for the *Wheel* as shown.

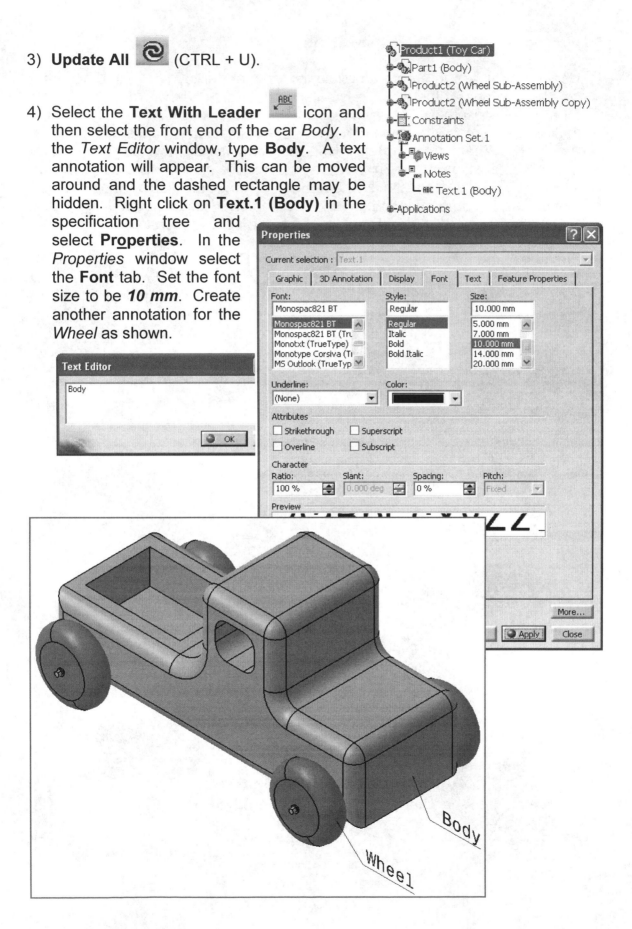

Chapter 5: ASSEMBLY DESIGN FUNDAMENTALS

Tutorial 5.2: Advanced Assembly

Featured Topics & Commands

Prerequisite Knowledge & Commands

- Entering workbenches
- The *Sketcher* workbench and associated commands
- The *Part Design* workbench and associated commands
- Moving and rotating parts with the compass

Tutorial 5.2 Start: Assembly/Parts Modeled

The assembly and parts modeled in this tutorial are shown in the figures below. The main objective of this tutorial is to familiarize you with the product structure and assembly constraints. However, assemblies are made up of parts that must fit together and fasten.   Therefore, the parts modeled in this tutorial have toleranced dimensions and threads.

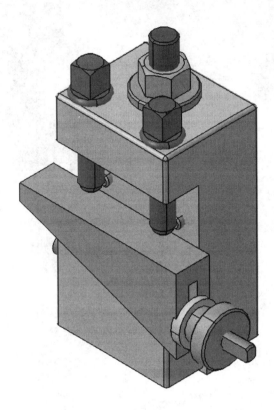

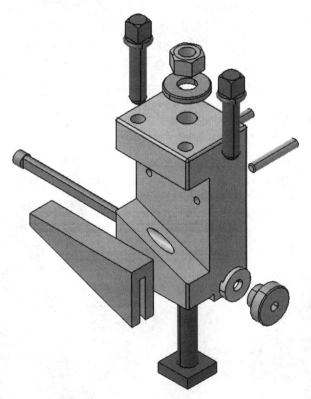

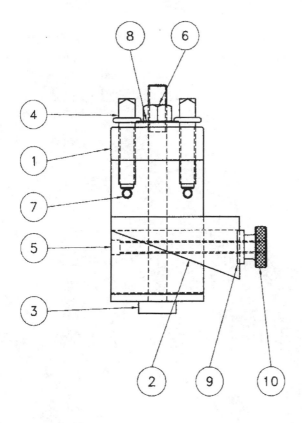

PART #	PART NAME	REQ'D	MAT
1	TOOL POST BODY	1	CI
2	WEDGE	1	SAE 1045
3	BOLT	1	SAE 1040
4	TOOL POST SCREW	2	SAE 1040
5	ADJUSTING SCREW	1	SAE 1040
6	HEX NUT	1	–
7	SET SCREW	2	SAE 1040
8	WASHER	1	SAE 1040
9	WASHER	1	SAE 1040
10	ADJUSTING NUT	1	SAE 1040

Section 1: Setting up your product

1) Enter the **Assembly Design** workbench. Notice that at the top of your specification tree there is a Product.

2) Set your units to be **Inches** (**Tools – Options...**).

3) Right click on **Product1** and select **Properties**. In the *Properties* window, click on the **Product** tab. In the Part Number field, rename your product *Tool Post*.

4) Save your draw as *Tool Post.CATProduct* in a folder called *Tool Post*.

Section 2: Part 1 (Tool Post Body)

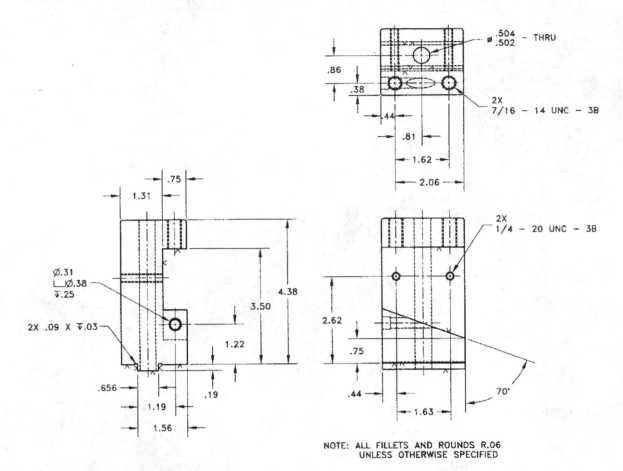

NOTE: ALL FILLETS AND ROUNDS R.06
UNLESS OTHERWISE SPECIFIED

1) Select the **Part** icon (If **Part1** does not appear in the specification tree, then click on **Tool Post**). This will add a new part to your assembly or product. Notice that *Tool Post* in the specification tree is highlighted in blue. This means that it is the active component.

2) Right click on **Part1 (Part1.1)** in the specification tree and select **Properties**. In the *Properties* window, click on the **Product** tab. In the Instance name field, rename your part **Tool Post Body**.

3) At the top pull down menu, select **File – Sa̲ve Management....** A *Save Management* window will appear indicating that *Tool Post.CATProduct* has been saved before, but modified since, and that *Part1.CATPart* is new. Click on **Tool Post.CATProduct** and select the **Save** button. Then, click on **Part1.CATPart** and select the **Save As...** button. Name your file **Part1 – Tool Post Body.CATPart**.

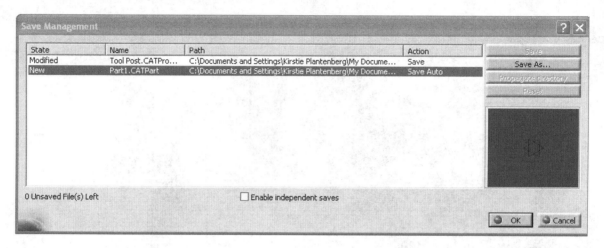

4) Expand *Part1 (Tool Post Body)* in the specification tree until you can see its *PartBody*. Double click on **PartBody**. This will automatically enter you into the *Part Design* workbench. You are now ready to draw Part1. Notice that now Part1 is highlighted in blue. If CATIA does not enter the *Part Design* workbench (it may enter the *Wireframe* workbench if this was the last one you were in), you will have to enter it manually.

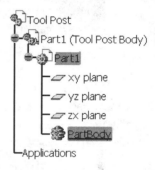

5) Draw the right side profile of Part1 on the **zx plane** as shown and *Pad* it to the appropriate length.

6) Fillet the appropriate edges.

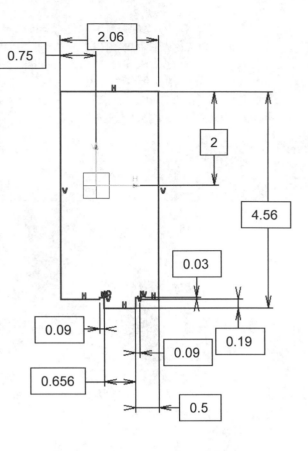

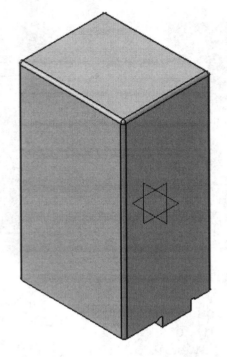

7) Draw and *Pocket* the slot in the front of Part1. Hint: Projecting the part edges will help.

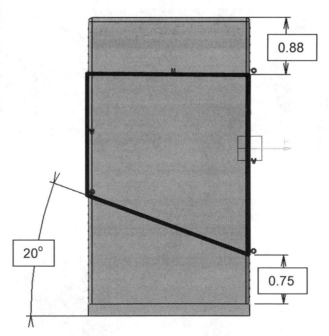

8) Create the *Hole* that travels from top to bottom. Notice that this hole is toleranced. Position the sketch before filling in the information. The command tends to lock up if you position your sketch at the end. To access the *Limit of Size Definition* window click on the tolerance icon. Note, the tolerance annotation shown on the CATIA part will be in metric (ϕ12.7 +0.1 to +0.05)

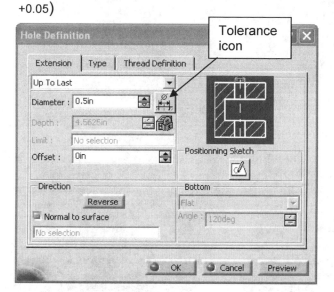

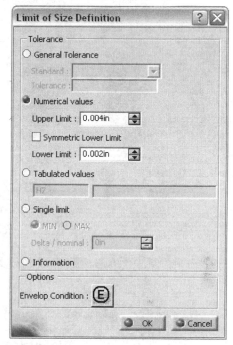

9) Create the two threaded *Holes* on the top. The thread note is 7/16 – 14 UNC – 3B. The Thread Diameter: (major diameter) is 7/16, the Hole Diameter: (tap drill) is 0.368 as obtained from a thread table, and the Pitch: is 1/14.

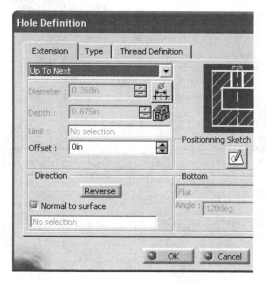

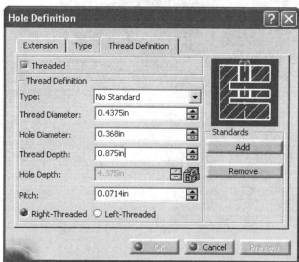

10)Create the two threaded *Holes* in the front.

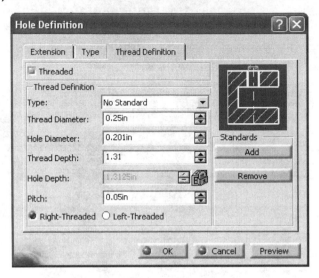

11)Create the Counter bored *Hole* in the side.

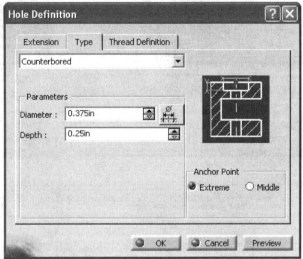

12) After completion, use **Save Management...** to save your product and part. Your part and specification tree should look like this.

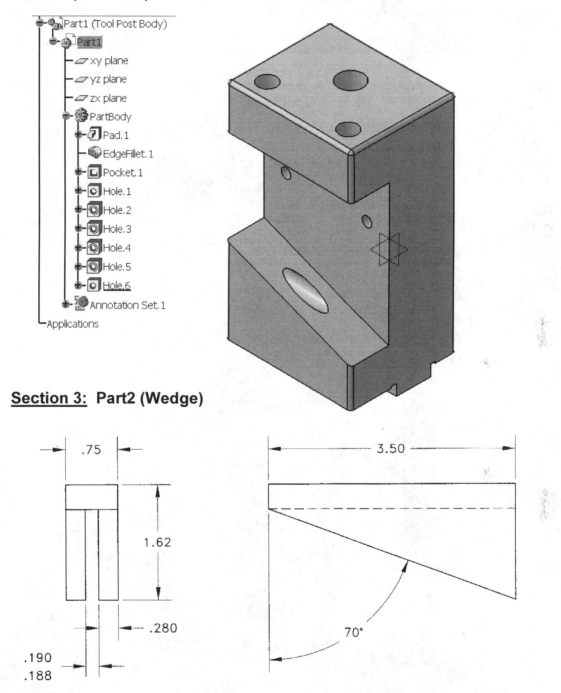

Section 3: Part2 (Wedge)

1) Enter the *Assembly Design* workbench by double clicking on **Tool Post**.

2) Select the **Part** icon and then click on **Tool Post** in your specification tree. You can select **No** in the *New Part: Origin Point* window.

3) Hide *Part1 (Tool Post Body)* by right clicking on it and selecting **Hide/Show**.

4) Rename your new part **Wedge**.

Tool Post
Part1 (Tool Post Body)
Part2 (Wedge)

5) Expand *Part2 (Wedge)* and double click on its **PartBody**. You are in the *Part Design* workbench and ready to draw the Part2.

6) To create a toleranced dimension, right click on the Value field in the *Constraint Definition* window and select **Add tolerance ....** When the *Tolerance* window appears, enter the max. and min. tolerance values.

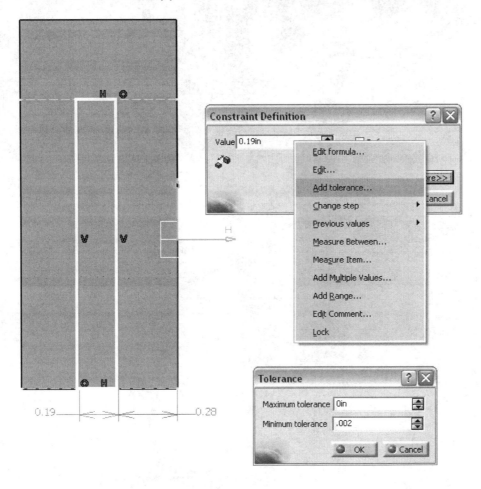

7) After completion, use **Sa<u>v</u>e Management...** to save you product and parts. Your part and specification tree should look like this.

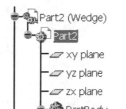

8) Enter the *Assembly Design* workbench by double clicking on **Tool Post**.

9) Unhide **Part1 (Tool Post Body)** and use the compass to move the *Wedge* away from the *Tool Post Body*.

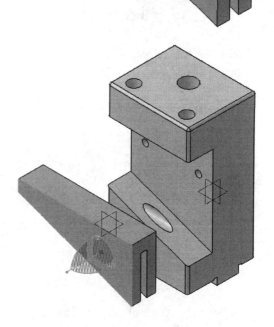

<u>Section 4:</u> Part 3 (Bolt)

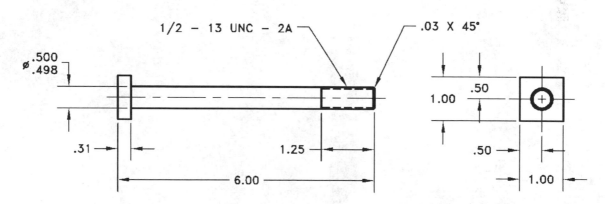

1) Select the **Part** icon and then click on **Tool Post** in your specification tree. You can select **<u>No</u>** in the *New Part: Origin Point* window.

2) Hide *Part1 (Tool Post Body)* and *Part2 (Wedge)*.

3) Rename your part **Bolt**.

4) Expand *Part3 (Bolt)* and double click on its **PartBody**. You are in the *Part Design* workbench and ready to draw the Part3.

5) Model the Bolt. Notice that the bolt shaft is toleranced. Use a basic size of 0.500 and the same procedure as before to apply the tolerance. The shaft also has threads. Use the

Thread/Tap command to apply them.

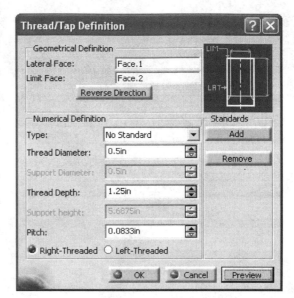

6) After completion, use **Sa̲ve Management...** to save you product and parts. Your part and specification tree should look like this.

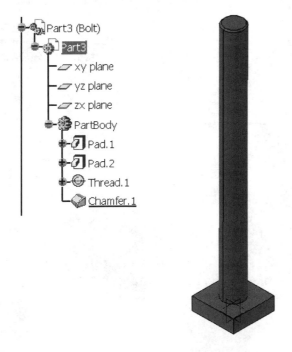

7) Enter the *Assembly Design* workbench by double clicking on **Tool Post**.

8) Unhide *Part1 (Tool Post Body)* and *Part2 (Wedge)*, and use the compass to move the *Bolt* away from the other parts.

9) **Save all**.

Section 5: Part4 through Part9

1) Model parts 4 through 9 and add them to the assembly as before. Note that parts 6, 7, and 8 are standard parts. CATIA contains a library of standard metric parts that can be accessed. However; this is an English drawing. Thus, we must draw the standard parts. Key steps for creating each part, the final specification tree, and final part are shown in a series of figures given below.

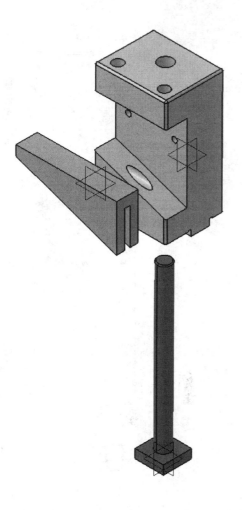

Part 4 (Tool Post Screw)

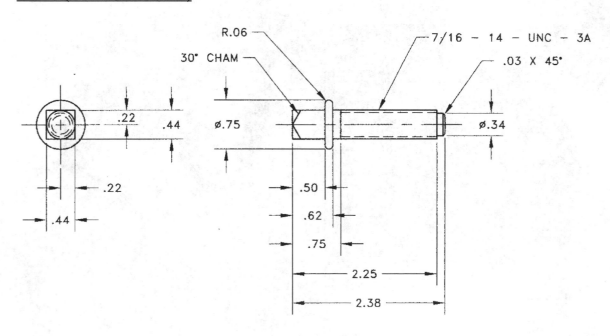

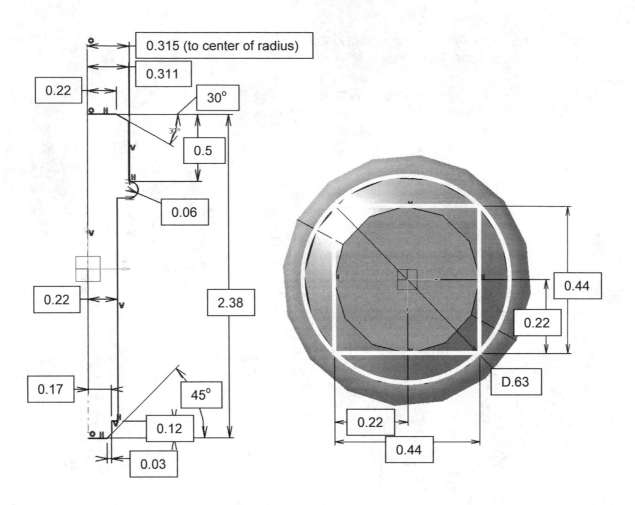

0.315 (to center of radius)

0.311

0.22

30°

0.5

0.06

0.22

2.38

0.17

45°

0.12

0.03

0.44

0.22

D.63

0.22

0.44

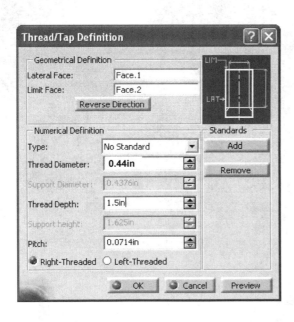

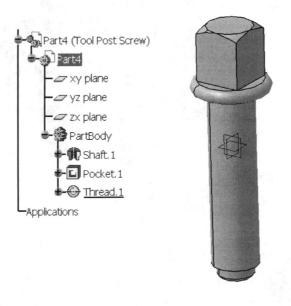

Part 5 (Adjusting Scew)

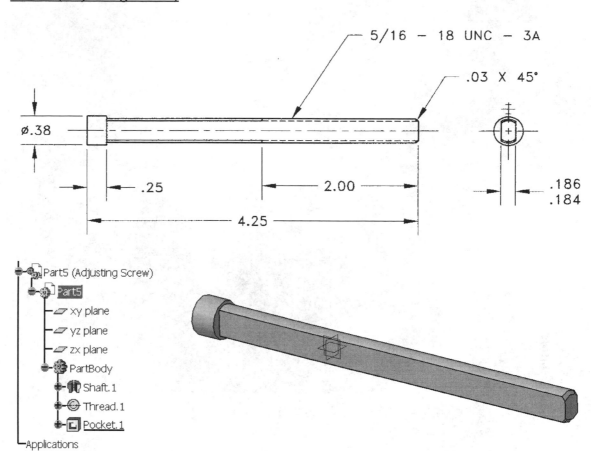

5/16 – 18 UNC – 3A

.03 X 45°

ø.38

.25

2.00

4.25

.186
.184

Part 6 (Hex Nut)

1/2 – 12 UNC – 3B

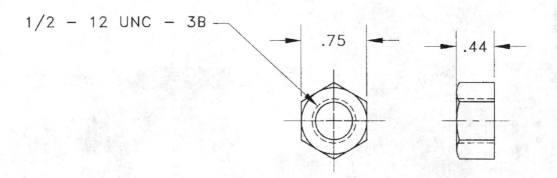

.75

.44

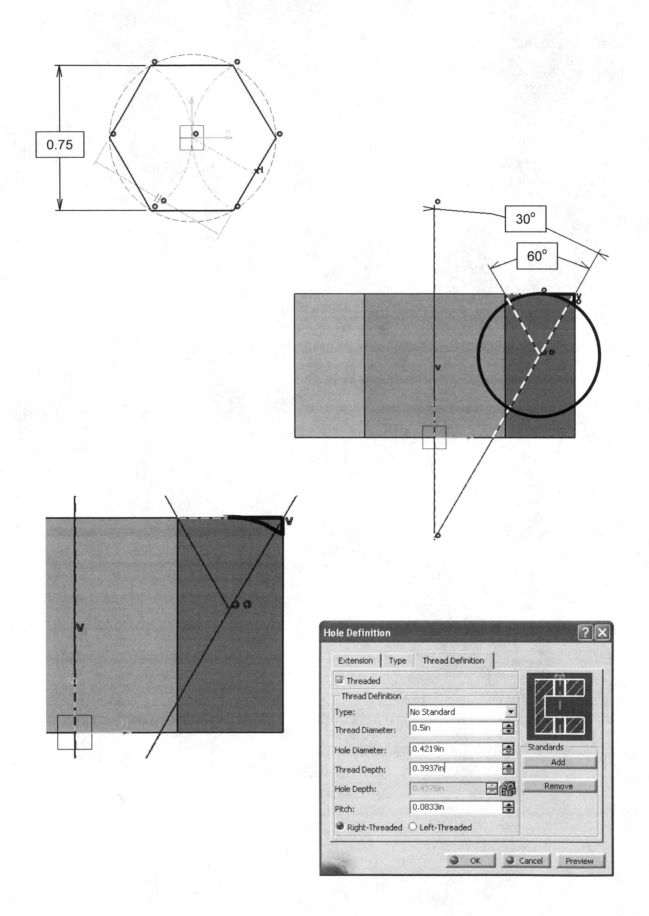

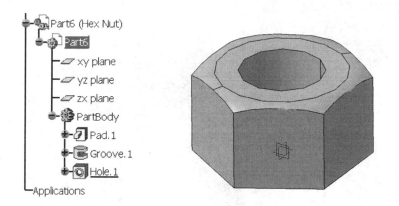

Part7 (Set Screw)

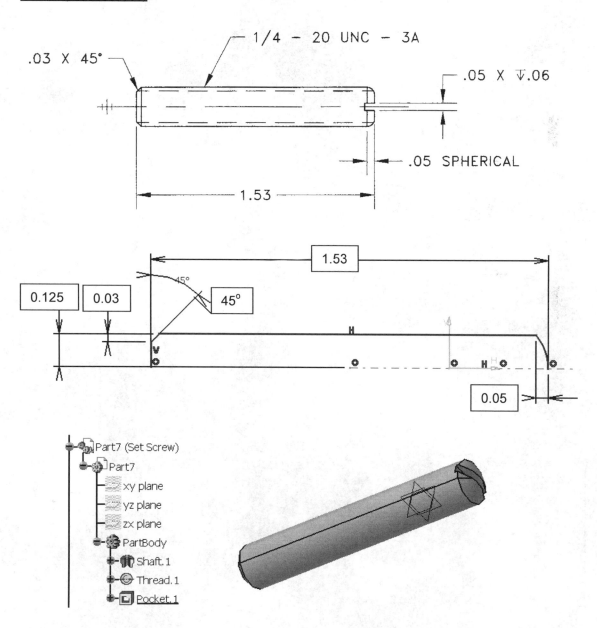

Part8 (Washer)

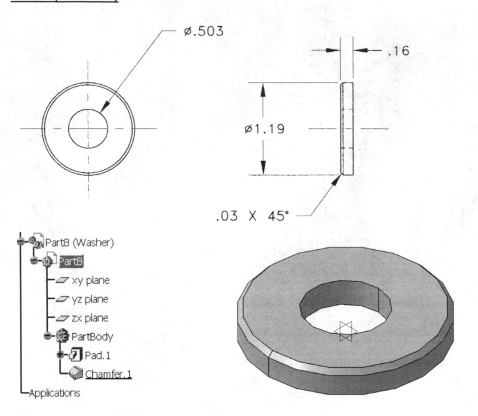

Ø.503

.16

Ø1.19

.03 X 45°

Part8 (Washer)
- Part8
  - xy plane
  - yz plane
  - zx plane
  - PartBody
    - Pad.1
    - Chamfer.1
- Applications

Part9 (Slotted Washer)

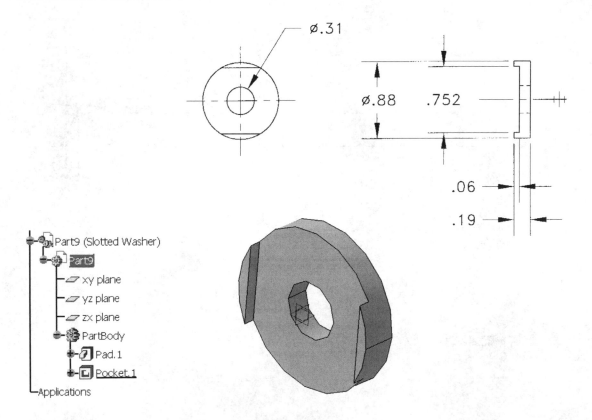

Ø.31

Ø.88   .752

.06

.19

Part9 (Slotted Washer)
- Part9
  - xy plane
  - yz plane
  - zx plane
  - PartBody
    - Pad.1
    - Pocket.1
- Applications

Part10 (Adjusting Nut)

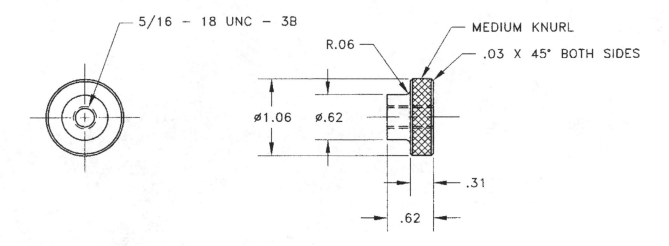

5/16 – 18 UNC – 3B

R.06

MEDIUM KNURL

.03 X 45° BOTH SIDES

ø1.06   ø.62

.31

.62

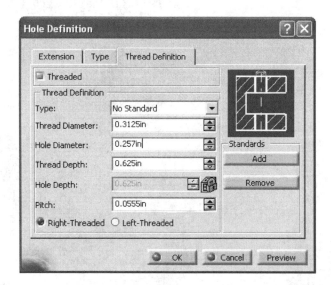

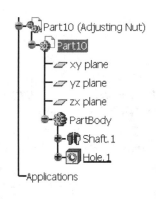

Section 6: Duplicating parts

1) After inserting and drawing all the parts, your specification tree should look like this.

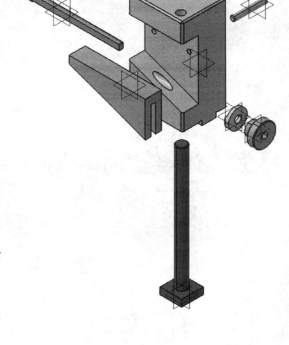

Tool Post
- Part1 (Tool Post Body)
- Part2 (Wedge)
- Part3 (Bolt)
- Part4 (Tool Post Screw)
- Part5 (Adjusting Screw)
- Part6 (Hex Nut)
- Part7 (Set Screw)
- Part8 (Washer)
- Part9 (Slotted Washer)
- Part10 (Adjusting Nut)
- Applications

2) Double click on Tool Post to enter the *Assembly Design* workbench.

3) Use your compass to move your parts in the approximate locations shown. To do this, grab the red square on the compass and move it to the part that you want to move. If the compass lines turn green, it has attached itself to the part. Use the lines and arcs to manipulate the part. When you are done, move it back to the upper right corner. It is now ready to attach it to another part.

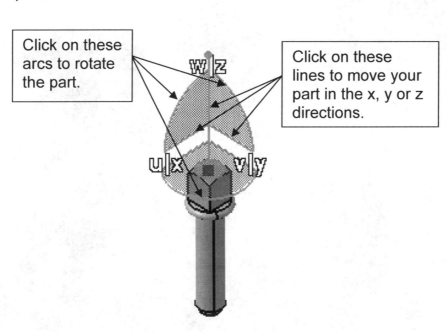

Click on these arcs to rotate the part.

Click on these lines to move your part in the x, y or z directions.

4) Notice that we are missing one *Tool Post Screw* and one *Set Screw*. To make copies, right click on the **Part4 (Tool Post Screw)** in the specification tree and select **Copy**. Then, right click on **Tool Post** (the root assembly) and select **Paste**. Repeat for the *Set Screw*. Name your new parts as shown in the specification tree.

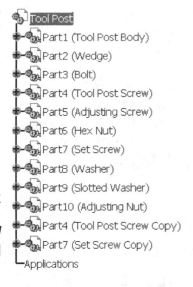

Section 7: Applying assembly constraints

1) Fix the *Tool Post Body*. Select the **Fix Component** 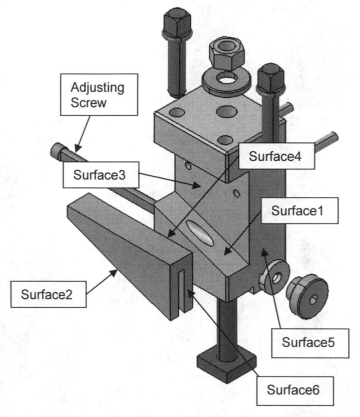 icon and then select the *Tool Post Body*. The *Fix Component* constraint symbol will appear on the *Tool Post Body*.

2) Apply a contact constraint between surface1 and surface2 as indicated in the figure. Select the **Contact Constraint** icon. If an *Assistant* window comes up, **Cancel** it. Select surface1 and rotate the assembly and select surface2 (the bottom surface of the wedge). A line with the contact constraint symbol at both ends will connect the two surfaces.

3) Apply a Contact Constraint between surface3 and surface4 (the back side of the wedge) as indicated in the figure.

4) Create an offset constraint between the surface5 and surface6. Select the **Offset Constraint** 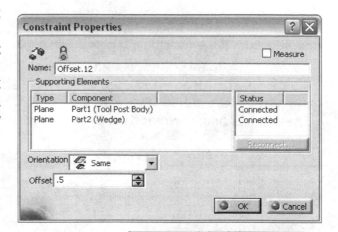 icon. Select surface5 and then surface6. A *Constraint Properties* window will appear. Fill in a value of *0. 5* inch in the Offset field.

5) The part is not in position yet. Select the **Update All** icon to move the part into position.

6) Apply a coincidence constraint between the *Adjusting Screw* and the counterbored hole in the *Tool Post Body*. Select the **Coincidence Constraint** icon. Select the axis of the *Adjusting Screw* (you may have to move the mouse around a bit or even zoom in before it shows up) and then the axis of the counterbored hole. A line with the coincidence constraint symbol at both ends will connect the two axes.

7) Apply a **Contact Constraint** between the bottom surface of the *Adjusting Screws* head (you will probably have to zoom in to be able to select it) and the bottom surface of the counterbored hole in the *Tool Post Body*.

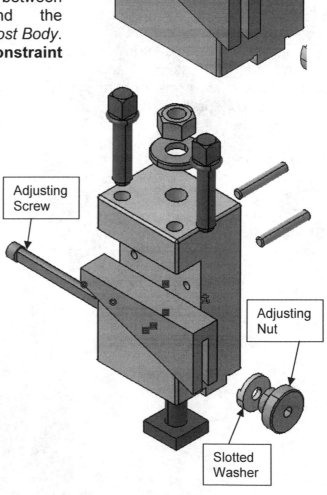

Adjusting Screw

Adjusting Nut

Slotted Washer

8) Apply the appropriate **Coincidence Constraints** 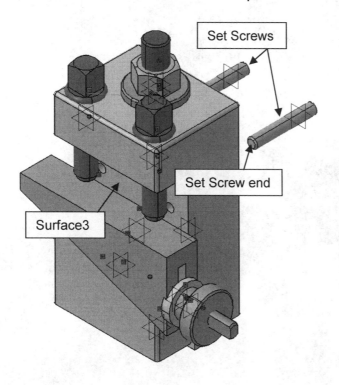 between the *Adjusting Screw*, and the *Slotted Washer* and *Adjusting Nut*. Remember to choose the axes and zoom in when necessary.

9) Apply the appropriate **Contact Constraints** between the *Wedge* and *Slotted Washer* and the *Slotted Washer* and the *Adjusting Nut*. Use the

Update All to move the parts to their proper positions.

10) Use the appropriate constraints to place the two Tool Post Screws, the Bolt, the Washer, and the Hex Nut into their functional positions.

11) Apply **Coincident Constraints** to the *Set Screws* and the holes in the *Tool Post Body*.

12) Create an offset constraint between the end of the *Set Screws* and surface3.

Select the **Offset Constraint** icon. Select the end of one *Set Screw* and then surface3. A *Constraint Properties* window will appear. Fill in a value of **-0.125** inch in the Offset field. Repeat for the other *Set Screw*.

Set Screws

Set Screw end

Surface3

13) Hide all the constraint symbols. Right click on **Constraints** in the specification tree and select **Hide/Show**.

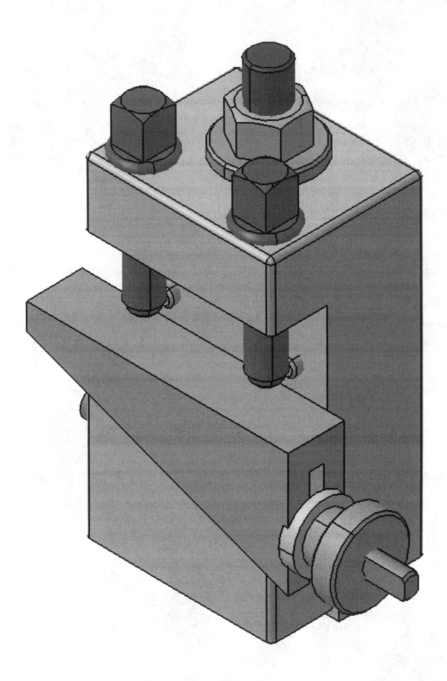

Chapter 5: Exercises

Exercise 5.1: This exercise can be performed after completing the tutorials presented in chapter 5. Model and constrain the following parts and assembly.

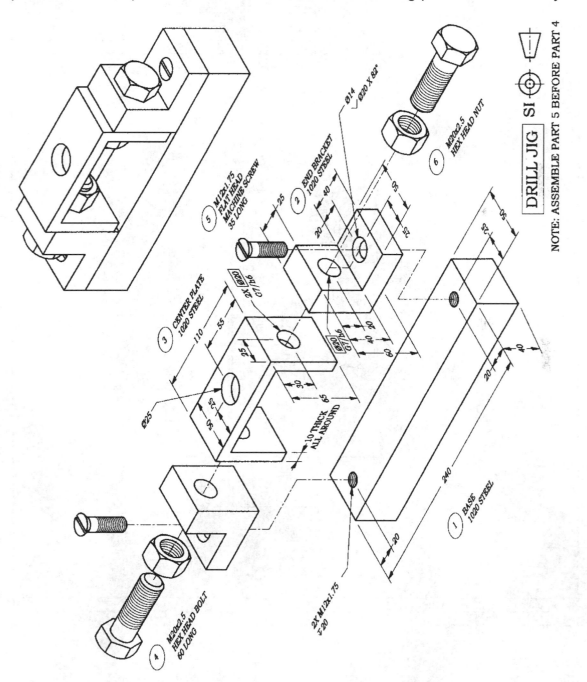

Exercise 5.2: This exercise can be performed after completing the tutorials presented in chapter 5. Model and constrain the following parts and assembly.

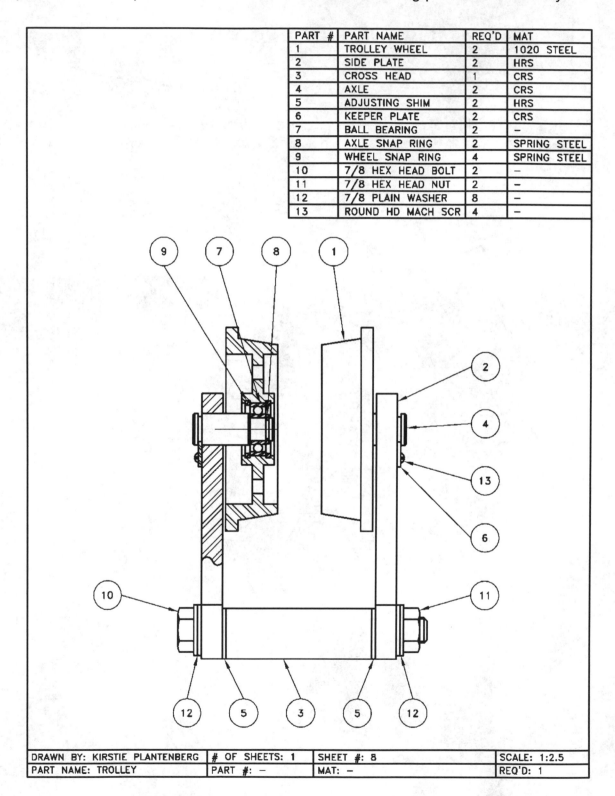

PART #	PART NAME	REQ'D	MAT
1	TROLLEY WHEEL	2	1020 STEEL
2	SIDE PLATE	2	HRS
3	CROSS HEAD	1	CRS
4	AXLE	2	CRS
5	ADJUSTING SHIM	2	HRS
6	KEEPER PLATE	2	CRS
7	BALL BEARING	2	–
8	AXLE SNAP RING	2	SPRING STEEL
9	WHEEL SNAP RING	4	SPRING STEEL
10	7/8 HEX HEAD BOLT	2	–
11	7/8 HEX HEAD NUT	2	–
12	7/8 PLAIN WASHER	8	–
13	ROUND HD MACH SCR	4	–

DRAWN BY: KIRSTIE PLANTENBERG	# OF SHEETS: 1	SHEET #: 8		SCALE: 1:2.5
PART NAME: TROLLEY	PART #: –	MAT: –		REQ'D: 1

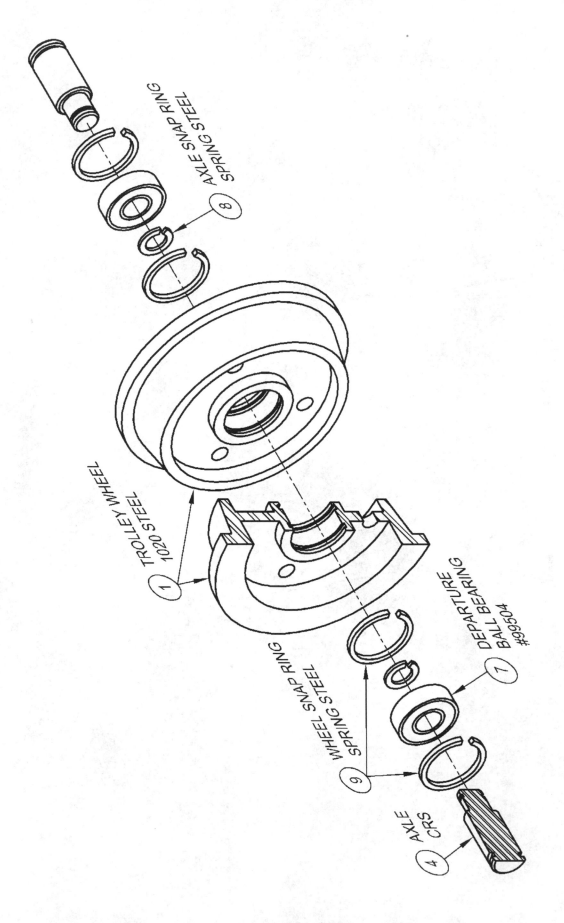

AXLE SNAP RING
SPRING STEEL
8

TROLLEY WHEEL
1020 STEEL
1

DEPARTURE
BALL BEARING
#99504
7

WHEEL SNAP RING
SPRING STEEL
9

AXLE
CRS
4

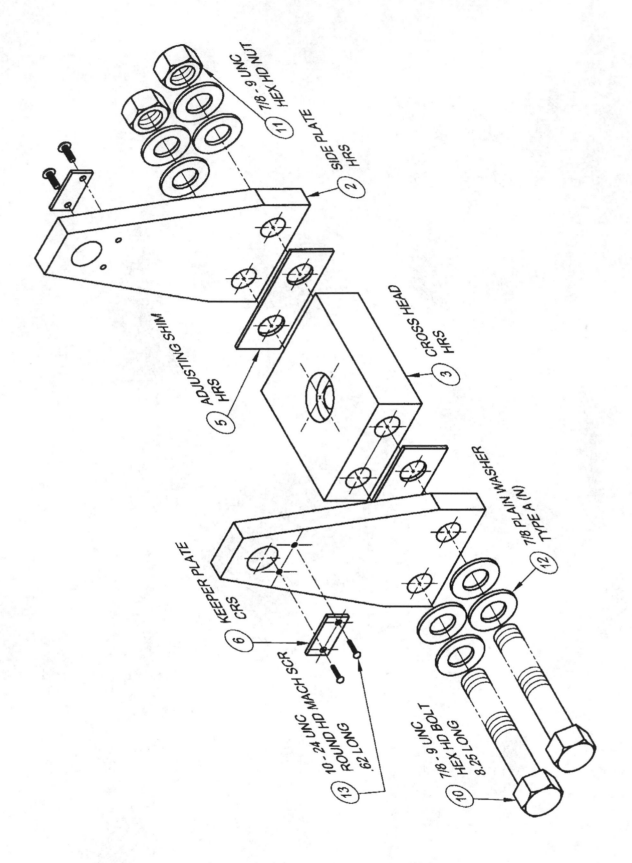

Part#1

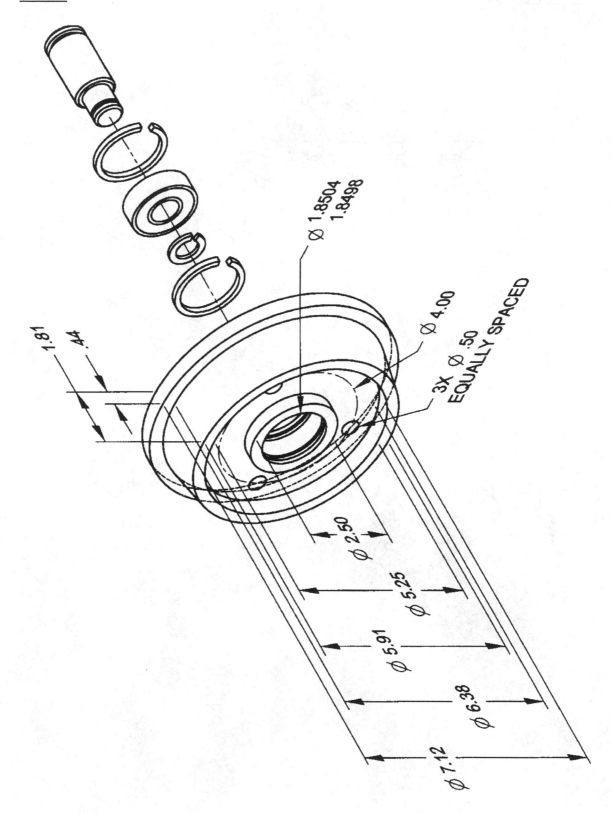

Ø 1.8504
1.8498

Ø 4.00

3X Ø .50
EQUALLY SPACED

1.81

.44

Ø 2.50

Ø 5.25

Ø 5.91

Ø 6.38

Ø 7.12

Part#1, Part#8 and Part#9

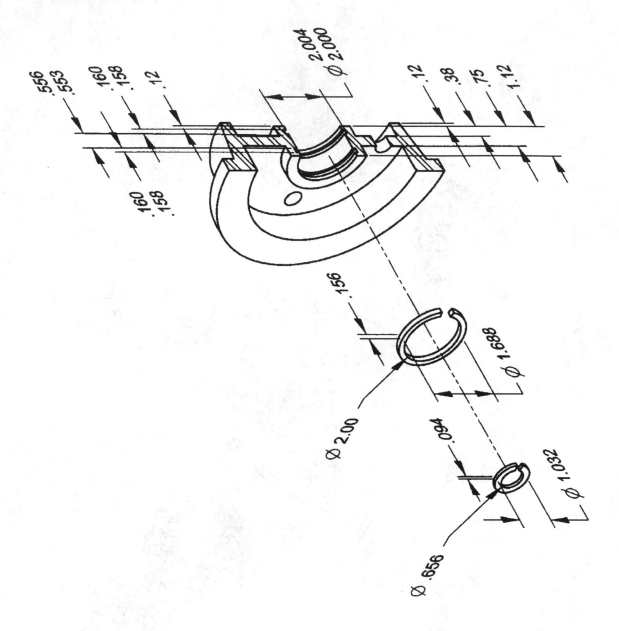

Part#4 and Part#7

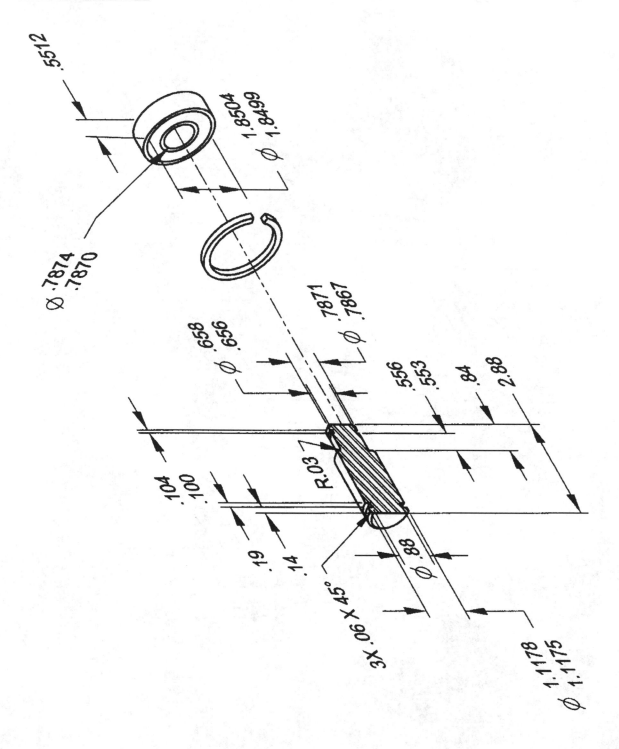

Part#2 and Part#6

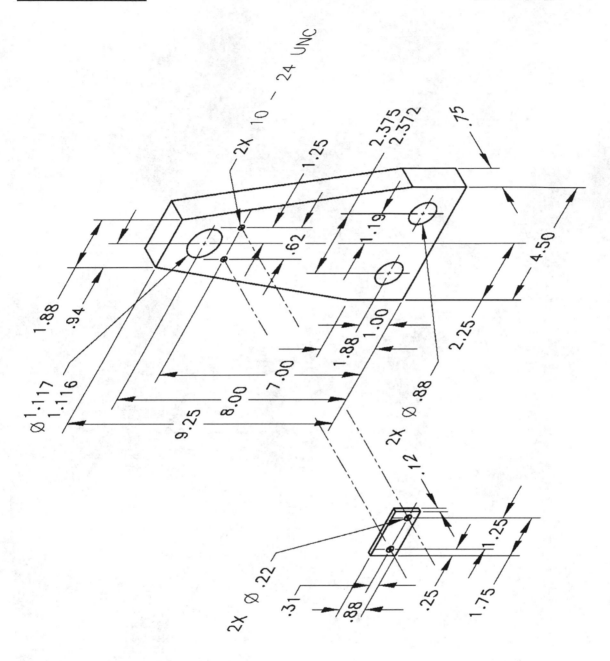

Part#3 and Part#5

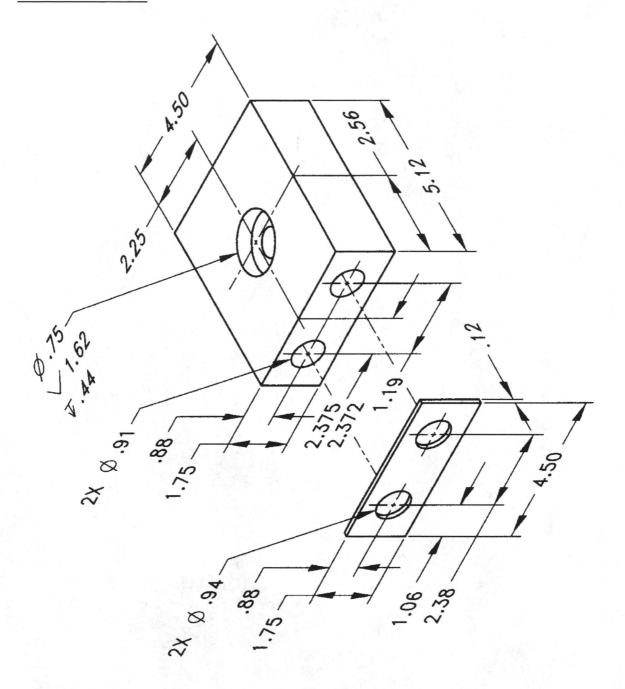

NOTES:

Chapter 6: DRAFTING FUNDAMENTALS

Introduction

Chapter 6 focuses on CATIA's *Drafting* workbench. The reader will learn how to create a simple orthographic projection and associated dimensions.

Tutorials Contained in Chapter 6

- Tutorial 6.1: Generative Drafting

NOTES:

NOTES:

Chapter 6:
DRAFTING FUNDAMENTALS

Tutorial 6.1: Generative Drafting

<u>Featured Topics & Commands</u>

<u>Prerequisite Knowledge & Commands</u>

- Entering workbenches
- The *Sketcher* workbench and associated commands
- The *Part Design* workbench and associated commands

The Drafting Workbench

The *Drafting Workbench* allows you to create an orthographic projection or drawing (CATDrawing) directly from a 3D part (CATPart) or assembly (CATProduct). A CATDrawing contains a structure listing similar to a specification tree. The structure listing shows all the sheets and views contained in the document. CATIA enables you to create *generative views* that are associative with the 3D part, and to create *drawn views* which are not associative.

The Views toolbar

The commands located in the *Views* toolbar enable you to create a variety of views and view configurations. The sub-toolbars located within the *Views* toolbar, reading left to right, are

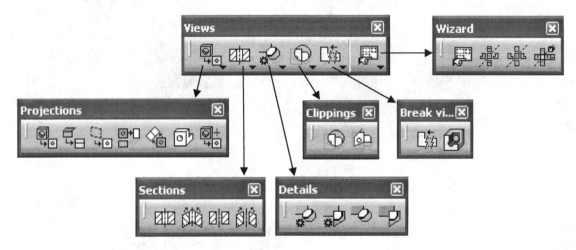

- Projections toolbar: The *Projections* toolbar contains commands that allow you to create different types of views. The commands, reading from left to right, are; *Front View, Unfolded View, View from 3D, Projection View, Auxiliary View, Isometric View, Advanced Front View*.
- Sections toolbar: This toolbar contains commands that allow you to create a variety of section and cut views.
- Details toolbar: The *Details* toolbar contains commands that allow you to create views that are a small portion of an existing view. The detail view is usually drawn at an increased scale.
- Clippings toolbar: The command located in the *Clippings* toolbar allow you to create removed views or removed section views. This is a small area of the part that is shown apart from the original view and is usually shown at an increased scale.

- <u>Break view toolbar:</u> The commands located in the *Break view* toolbar allow you to break the part in a specified location. This is usually done to save drawing space. The missing section of the part is usually uninteresting and not worth showing. There is also a command that enables you to create a broken out section.

- <u>Wizard toolbar:</u> This toolbar contains commands that allow you to select from several predefined view configurations or to define your own custom configuration. Reading from left to right, the commands are; '*View Creation Wizard*', which allows you to create custom view configurations, '*Front, Top and Left*', '*Front, Bottom and Right*', and '*All Views*'.

The Dimensioning toolbar

The commands located in the *Dimensioning* toolbar allow you to manually create dimensions and tolerances. The sub-toolbars, reading from left to right, are

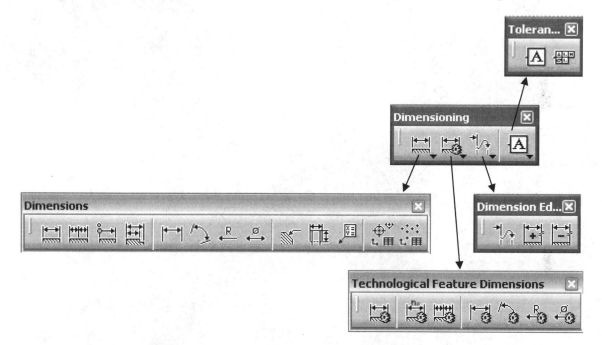

- <u>Dimensions toolbar:</u> The *Dimensions* toolbar contains commands that allow you to manually create dimensions. The commands, reading from left to right, are; *Dimensions, Chained Dimensions, Cumulated Dimensions, Stacked Dimensions, Length/Distance Dimensions, Angle Dimensions, Radius Dimensions, Diameter Dimensions, Chamfer Dimensions, Thread Dimensions, Coordinate Dimensions, Hole Dimension Table,* and *Coordinate Dimension Table*.

- <u>Technological Feature Dimensions toolbar:</u> These commands allow you to dimension technological features. Technological feature dimensioning relies on the fact that technological features can specify the way they should be dimensioned. This allows you to create only realistic and customized dimensions, based on the know-how of a given field.

- <u>Dimension Editor toolbar:</u> The commands in this toolbar allow you to re-route dimensions, and create or eliminate interruptions.
- <u>Tolerancing toolbar:</u> The Tolerancing toolbar contains commands that allow you identify Datum features and to apply feature control frames with associated GD&T symbols.

Part/Drawing Modeled

The part modeled in this tutorial is used to illustrate the basic commands available in the *Drafting* workbench. You will learn how to create an orthographic projection that contains a section view, and how to create and place dimensions.

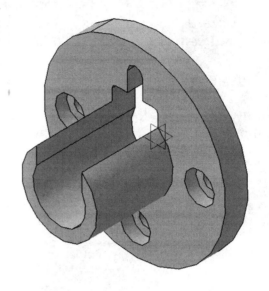

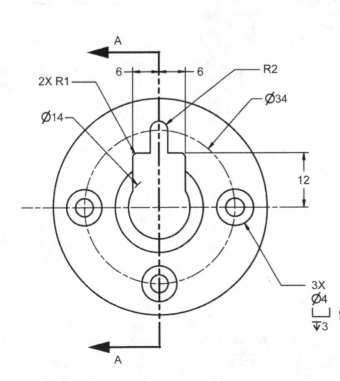

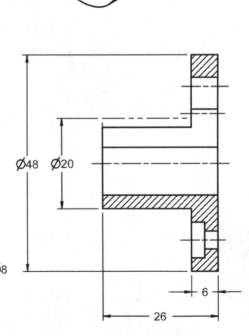

Section 1: Modeling the part

1) Open a **New ...** Part, name it **Jig** and save the file as **T6-1.CATPart**.

2) Model the part shown on the previous page in the *Part Design* workbench and save.

Section 2: Creating standard views

1) Make sure your part is in a similar position as shown in the *Part/Drawing Modeled* section. Your isometric view is based on the orientation of you solid model.

2) With the **PartBody** of your part selected switch to the *Drafting* workbench. In the *New Drawing Creation* window, select **Empty sheet** layout and then **OK**. You can select the *Modify...* button if you need to change the paper size, but the A size paper will work for this part.

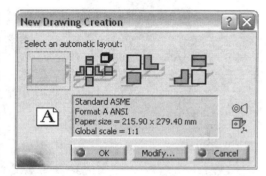

3) When the blank sheet appears, right click on **Sheet.1** in your structure listing (on the left) and select **Properties**. In the Properties window make sure that the **third angle standard** is active.

4) Select the **View Creation Wizard** icon. In the *View Wizard (Step 1/2) Predefined Configuration* window, select the **Front, Top and Right** configuration icon. Select the **Next >** button. In the *View Wizard (Step 2/2): Arranging the Configuration* window, select the **Isometric view** icon and place the view as shown. Select the **Finish** button.

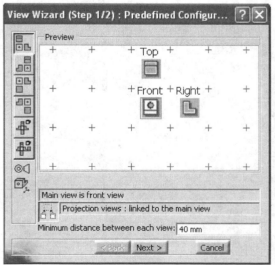
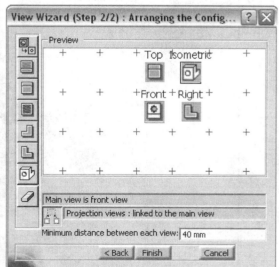

5) The prompt line will read, *Select a reference plane on a 3D geometry*. At the top pull down menu, select **Window** and then your part drawing. Move your mouse over the front face of the part. An *Oriented Preview* window will appear in the corner showing you what the front view will look like. If this is acceptable, click the mouse.

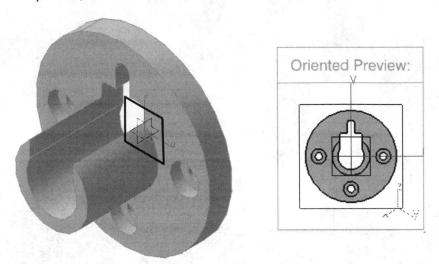

6) Your sheet should contain the front view and areas for the top, right and isometric views. These are not necessary in their final positions. The prompt line will read, *Click on the sheet to generate the view or redefine the view orientation using the arrows*. Look at the front view. If the orientation is not correct, you can use the blue circle with four arrows to reorient the front view. If the orientation is correct, click on the sheet.

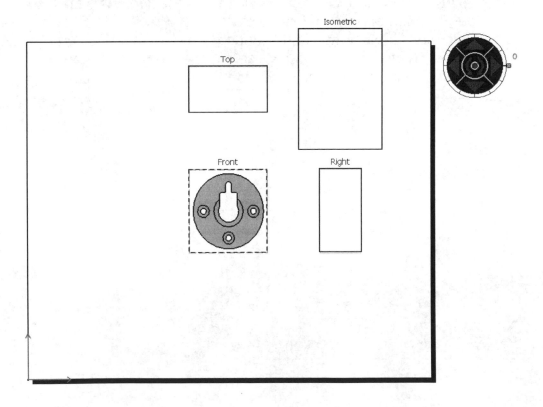

7) After you accept the front view orientation by clicking on the screen, the other views will appear. The front view should be the view that is active (in a red border). If it is not active, double click on it. Move the front view towards the bottom left corner, and then move the other views as shown in the figure. To move a view, click on the view border and move the mouse. Notice that when you try to move the isometric view it is constrained by the location of the front view. Right click on the isometric view's border and select **<u>V</u>iew Positioning – <u>P</u>osition Independently of Reference View**. Then, move the isometric view to the position shown.

(**Problem?** If you have no view borders, activate the *Display View Frame as Specified for Each View* icon. It is usually located in the bottom toolbar area.)

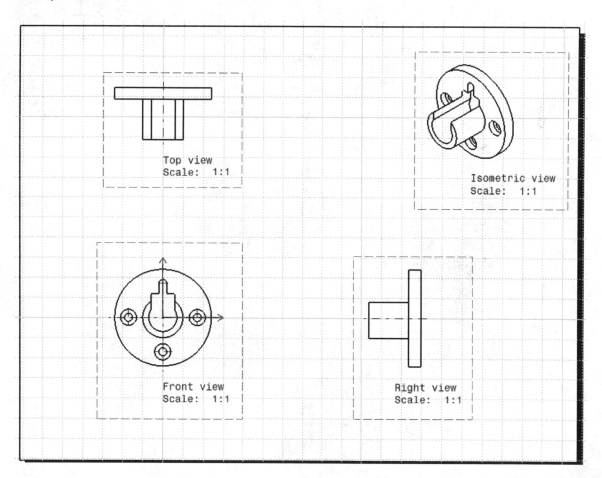

8) **Save** your drawing as *T6-1.CATDrawing*.

9) Notice that none of the views contain hidden or center lines. Right click on the **Right view** identifier in the structure listing and select P**r**operties. In the *Properties* window, activate the **Hidden Lines, Center Lines,** and **Axis** toggles in the *Dress-up* area. Repeat for the front and top views.

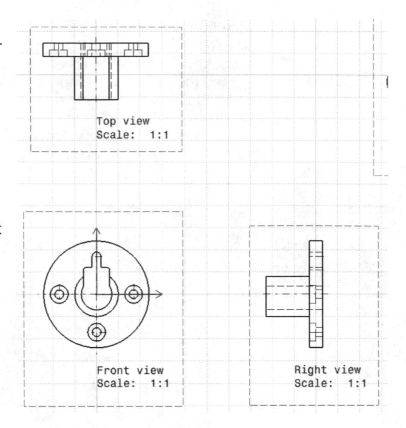

Top view
Scale: 1:1

Front view
Scale: 1:1

Right view
Scale: 1:1

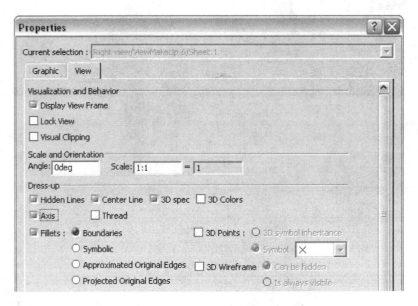

Section 3: Creating section views

1) Create a full section right side view. Select the **Offset Section** icon. Choose the location of your cutting plane in your front view by selecting point1 as indicated in the figure and then double clicking at point2. A shaded view will appear. Place it to the right of the right side view and click.

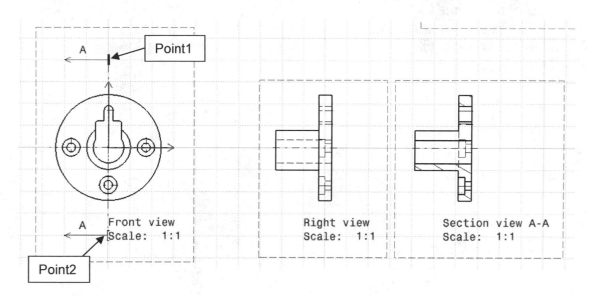

2) Change the look of the section lines. In the section view, double click on the section lines. In the *Properties* window click on the **Pattern** tab and change the angle to **45°** and the pitch to **1 mm**.

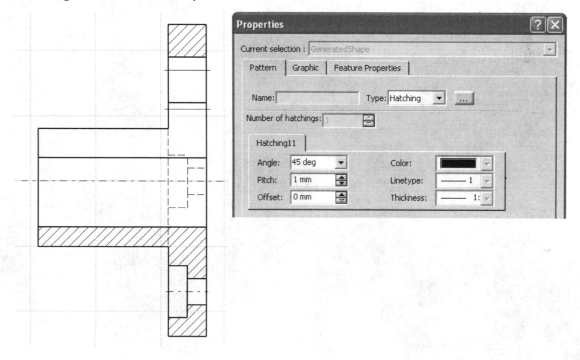

Section 4: Cleaning up the views

1) Delete the right side unsectioned view and the top view for these views are not needed. Right click on the border of the view and select **Delete**.

2) Notice that in the right side sectioned view there are two places where two lines (center and hidden) are coincident. According to drafting standards only one of those lines should be shown. Usually the center line would be deleted, but in this case the hidden lines have no reason for being there (they are used to show a fillet).   Enter the *Properties* window for the right side view. Deselect the **Fillet** toggle.

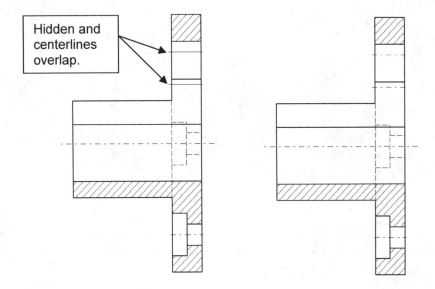

Hidden and centerlines overlap.

3) Section views usually do not contain hidden lines. Eliminate the hidden lines in the right side section view.

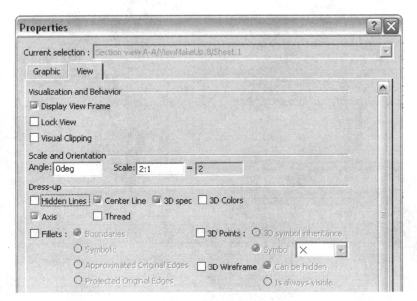

4) Change the properties of the cutting plane line. Right click on the cutting plane line and select **Properties**. In the *Properties* window, click on the **Call out** tab and set the line thickness to *3* and the line type to *5*. Set the length of the arrow base to *0.001 mm*. Set the arrow head type to solid, the length to *5 mm*, and the angle to *30°*.

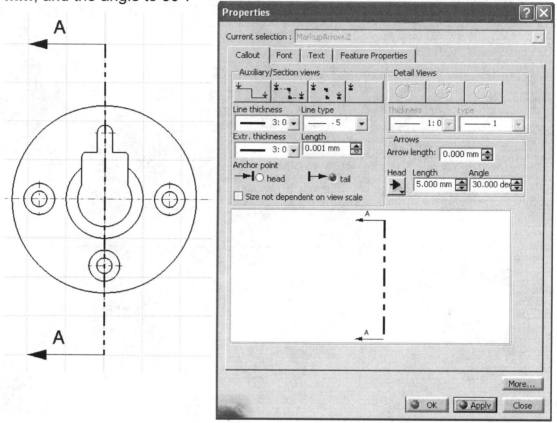

5) Hide the view names. Right click on each view name and select **Hide/Show**.

6) Change the scale of the views to 2:1. Right click on Sheet.1 and select **Properties**. In the *Properties* window set the scale to *2:1*.

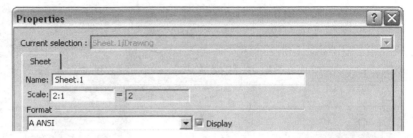

7) Change the scale of the isometric view back to **1:1**.

8) Save your drawing.

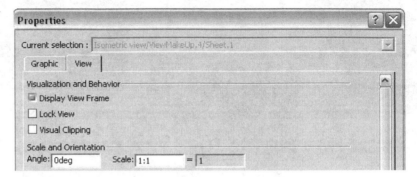

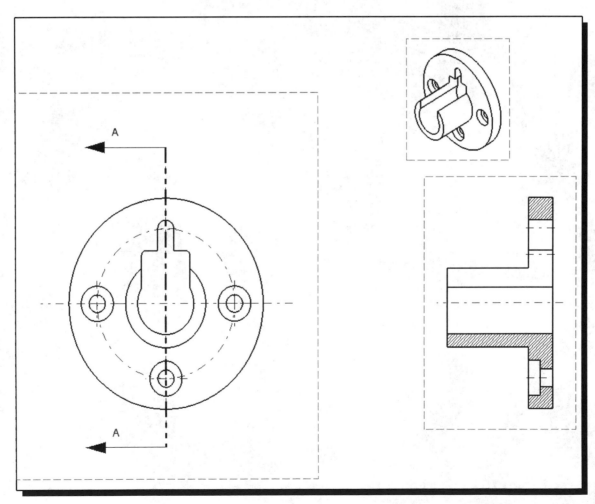

Section 5: Dimensioning

1) Make the right side view active by double clicking on its frame/border.

2) Draw a phantom line that indicates the full diameter of the hub. Draw two lines as indicated in the figure. Select both lines, right click, and select **Properties**. In the *Properties* window, set the line thickness to **1** and the line type to **5**. Commands such as line, circle, etc... are located in the *Geometry Creation* toolbar. (Hint: Start the vertical line at the centerline so that you know how long to make it.)

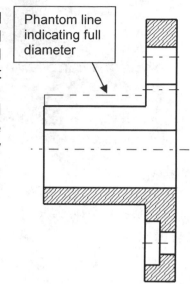

3) Place the linear dimension in the right side view. Select the **Length/Distance Dimensions** icon. Look up at the top and find the *Numerical Properties* toolbar. It should read **NUM.DIMM**. That means that your dimension values will be in millimeters. Then, select *Line1* followed by *Line2*. Repeat, using a similar procedure, for the 26 mm dimension. Note: You can move the arrows in and outside of the extension lines by clicking on them. You can also drag the texts outside the extension lines.

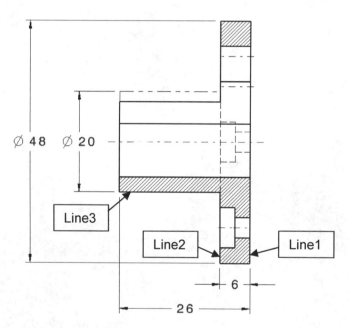

4) Select the **Diameter Dimension** ![diameter dimension icon] icon. Select *Line3* (See previous figure). The ⌀20 mm diameter dimension should appear. If it does not, select the horizontal phantom line. Repeat for the ⌀48 mm diameter dimension.

5) Activate the front view by double clicking on it.

6) Draw a **Circle** ![circle icon] that identifies the circle of centers of the counterbored holes.

7) Right click on the circle and select **Properties**. Change the line type to *4* and the line thickness to *1*.

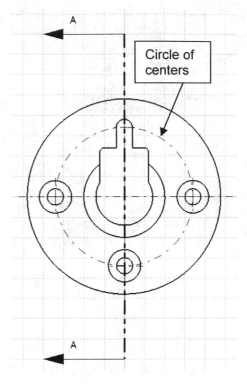

Circle of centers

8) Dimension the 3 **Diameter Dimensions** 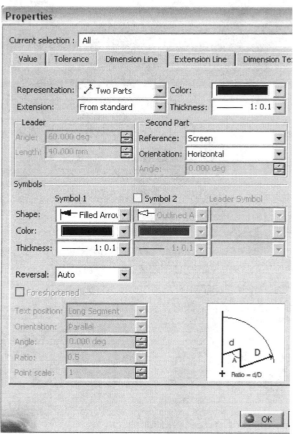 shown in the figure. Notice that the dimensions do not look like the dimensions shown in the figure. Select all 3 dimensions holding the Ctrl key and access the *Properties* window. In the **Dimension Line** tab change the following settings.

- `Representation`: **Two Parts**
- Deactivate the `Symbol 2` toggle.

Position the dimensions as shown.

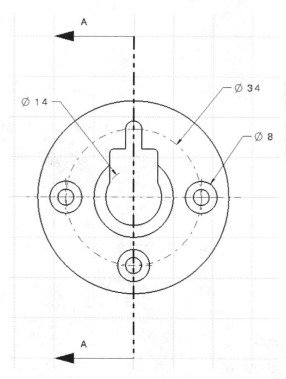

9) Notice that the dimension for the counterbored hole is not correctly stated. It shows only the diameter of the counterbore. Access the *Properties* window for that dimension. Select the **Value** tab and activate the **Fake Dimension** toggle. Enter a Main Value: of **4 mm**. This is the diameter of the hole. Select the **Dimension Text** tab. Enter **3X** in the top box of the Main Value field. This indicates that there are 3 counterbored holes.

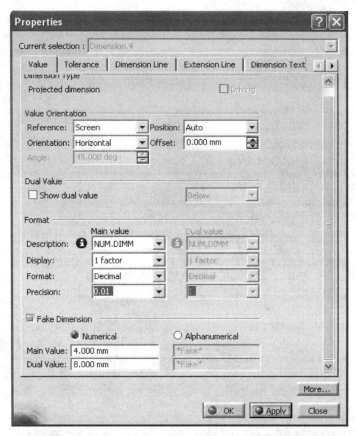

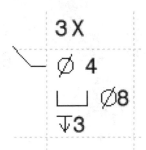

10) We still need to add the counterbore diameter and depth. Select the **Text** 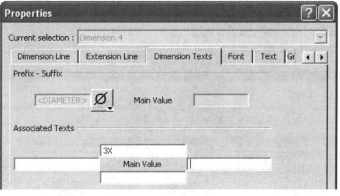 icon. Click your mouse under the ⌀4 dimension. A *Text Editor* window will appear. At the top of the screen there should be a *Text Properties* toolbar. Click on the **Insert Symbol** icon and select the counterbore ⎍ symbol, type a space, select the diameter symbol ⌀ , and then type **8**. Hit **Shift-Enter** to start a new line. Select the deep ⊤ symbol, and then type **3**.

11) Move the counterbore dimension and text down as shown in the figure. Hold the `Shift` key down while moving the counterbore text to avoid a jerky motion.

12) **Save** your drawing.

13) Dimension the rest of the linear features using **Length/Distance Dimensions** 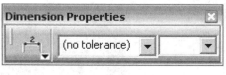.

14) Dimension the radius dimensions using the **Radius Dimension** 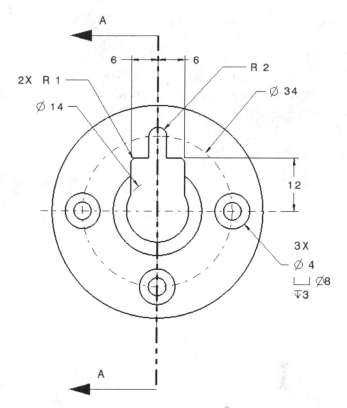 commands. Notice that the radial dimension don't look like what is shown in the figure. At the top of the screen there should be a *Dimension Properties* window. Select one of the radial dimensions and select the **Dimension Line** 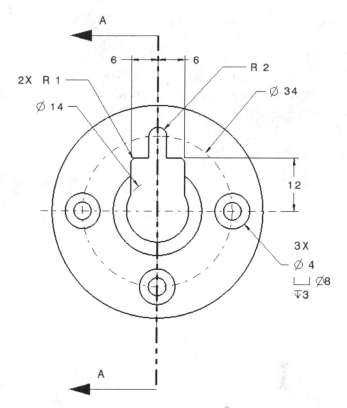 configuration. Right click on the R1 dimension and select **Pro\u005fperties**. In the *Properties* window, select the **Dimension Texts** tab and enter **2X** in the left box of the `Main Value` field.

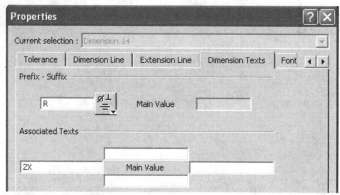

15) Position your views and then select the **Display View Frame as Specified**

for Each View 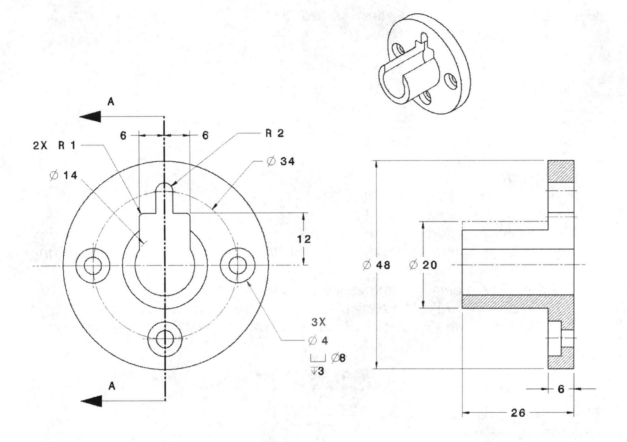 icon. This should turn off all of the view frames. This icon is usually located in the bottom toolbar area. If you are having trouble locating this icon, access the *Properties* window for each view and deactivate the **Display View Frame** toggle.

16) **Save** your drawing.

Chapter 6: Exercises

Exercise 6.1: This exercise can be performed after completing the tutorials presented in chapter 6. Create a detailed drawing of the exercises presented in chapter 2 and 3. The detailed drawing should include the appropriate views as well as an isometric view. Include section views were necessary. Dimension the orthographic projection completely.

NOTES:

NOTES:

NOTES:

NOTES:

NOTES: